Television Performing

ELECTRONIC MEDIA GUIDES

Elizabeth Czech-Beckerman *Managing Electronic Media*

Ken Dancyger *Broadcast Writing: Dramas, Comedies, and Documentaries*

Donna L. Halper *Full-Service Radio: Programming for the Community*

Donna L. Halper *Radio Music Directing*

William Hawes *Television Performing: News and Information*

Robert L. Hilliard *The Federal Communications Commission: A Primer*

John R. Hitchcock *Sportscasting*

Andrew F. Inglis *Satellite Technology : An Introduction*

Carla B. Johnston *Election Coverage: Blueprint for Broadcasters*

Marilyn J. Matelski *TV News Ethics*

Marilyn J. Matelski *Daytime Television Programming*

Robert B. Musburger *Electronic News Gathering: A Guide to ENG*

Television Performing

News and Information

William Hawes

University of Houston

Focal Press

Boston London

To Nancy Hawes Wagner

Focal Press is an imprint of Butterworth–Heinemann.

Recognizing the importance of preserving what has been written, it is the policy of Butterworth–Heinemann to have the books it publishes printed on acid-free paper, and we exert our best efforts to that end.

Library of Congress Cataloging-in-Publication Data
Hawes, William, 1931-
Television performing: news and information / William Hawes.
p. cm. — (Electronic media guide)
Includes bibliographical references.
ISBN 0-240-80056-7 (paperback)
1. Television broadcasting of news—Psychological aspects.
2. Television broadcasting of news—Public opinion. 3. Journalists—Attitude.
4. Self-presentation. 5. Public relations.
6. Personality. 7. Creative ability. I. Title. II. Series.
PN4888.T4H39 1991
791.45'028—dc20 90-40737
CIP

British Library Cataloguing in Publication Data
Hawes, William
Television performing: news and information. – (Electronic media guide)
1. Television programmes. News programmes. Journalism
I. Title II. Series
070.19

ISBN 0-240-80056-7

Butterworth–Heinemann
80 Montvale Avenue
Stoneham, MA 02180

10 9 8 7 6 5 4 3 2 1

Printed in the United States of America

Contents

Preface

These days just about anyone who wants to be on television can be, at least for a short time. If you appear on television, you become a performer. You and your viewers instantly begin to assess each other.

Performers fall mainly into two categories: performers who reveal news and information about the world and its inhabitants, and performers who are entertainers. At the outset of your career you decide whether you are a reporter of news and information or an entertainer. Fundamentally, an information person tells about other people and events, and an entertainer for the most part *is* the event. There is a sanctity and responsibility about journalism that prohibits a reporter from considering himself an entertainer. Those in entertainment do not want to give an account of the lives of other people, they generally want to be other people or characters. The truth they seek is in the spirit of what they do: They are artists. By contrast, the journalists seek truth in the literal description of persons and events from a factual perspective.

How do you decide whether you belong in information or entertainment? The answer is, intuitively. You already know your preference. Even though you like both information and entertainment to some extent, here is the test: If you could do anything you want in the world of performance, what would you do? Set aside the advice someone else (who may never have been a performer) has given you, and set aside the obstacles, real and imagined, in obtaining your goal. Now ask yourself what you really want to do as a performer during your lifetime. Your decision will be based on heartfelt desire, background, aptitude, perception of opportunity, advice from others, and whether you are willing to pay the price for being a performer. Keep in mind that being a performer can be a highly beneficial beginning to your life and career. The experiences you have in attempting to do so may lead to other kinds of professional opportunities in later life. A reasonable way of approaching performing is to set short term goals that are realistic. Nevertheless, if you want to be a performer, then you must make a commitment to it. Few reporters are part-time and few actors are part-time *in their commitment*. Their employment may only be sporadic, but their commitment is full-time. If you cannot make such a commitment, or if you vacillate, or if you do not intuitively *know* you are a performer, chances are you aren't. You may appear as a performer occasionally, but your life's work will probably take some other direction.

This book focuses upon the journalist or information specialist. These roles include three groups: (1) the career professional, (2) the community expert, and (3) the private citizen who, like it or not, is thrust into the media spotlight. Some career possibilities are evident. They include news anchors, reporters, sportscasters, and weathercasters; feature producers on health care, economics, government, gardening, cooking, business, or personal problems; consumer advocates; hosts for community affairs programs, documentaries, and fundraisers; narrators, salespersons, and instructors. All of these performers expect to be paid by a television company for their services. They are in the business of selling information.

By contrast, the community expert may or may not be paid directly by a television company. Instead, the minister, physician, attorney, stock analyst, educator, pollster, literary, or entertainment figure will probably be compensated by the business or government institution that hires a person principally for the person's knowledge of the subject. Television exposure is secondary to another kind of valuable service. For experts, television appearances recognize and enhance their roles in life.

A third type of individual appears on television as much by chance as by design. This person is on television because of having waved to mom while visiting a flower show being videotaped on location. This child stood in front of the prize roses. Or this individual may have been a witness to an event, such as a shooting or a fire, that incurred public intervention. More likely, this performer will be on television because the individual has purchased television equipment and videotaped himself or herself. Regardless of the circumstances, no matter what anyone utters aloud, almost everyone who appears on television for any reason wants to look good, sound intelligent, and be admired.

For you to be effective on television, and for you to maximize the benefits of an appearance as an information specialist, there are many considerations. Whether you are a career journalist, an information expert, or a private citizen, these roles require forethought and skill. So, in an information society, where television is the dominant medium, what you attempt to do is to control your image to the extent that in every circumstance you seem to be at your best. This book should be a helpful guide in reaching your objectives as a television performer in the increasingly complex world of news and information.

1 The Material

If you are going to be a performer, then you must have something to do, something to perform. This is called your *material*. Your material may be self-generated, original, and, therefore, self-owned; or it may be written by others, and they may own it. You or the company you work for would pay a fee to use it. Broadly speaking, material for performers is mainly news and information or entertainment. Of course, news can be entertaining at times, and entertainment can be newsworthy or informative at times, but the material is written to be one or the other.

Your material is fundamental to who you are and what you do. Your material, for example, may center on reporting current events. You may be perceived as a world traveler, sophisticated, experienced, an acquaintance of world leaders, and an eyewitness to political change, human sorrow, and triumph. This image may be different from that of a local news reporter who dependably covers those stories impacting a small community, but the changes and human problems locally are just as important to those people as they are to others on the world scene. Significance is measured by doing a good job wherever you are. Your value as a performer grows from the quality of your material. Having decided that you are a journalist or information specialist, you must look for sources of material.

SOURCES OF MATERIAL

For the information specialist the sources of material are virtually the entire range of human experience. Traditionally, you may think of newscasts, documentaries, and informational programs as the types most commonly broadcast. Talk, advice, instruction, ministries, product orientation, and sales are other categories on cable and cassette that utilize performers. The information age offers abundant opportunity for every performer who has expertise to share. For journalists the common sources of information are (1) being an eyewitness, that is, you saw the tornado touch down, you were in the war; (2) gathering information at the scene from participants, such as those involved in the accident; (3) getting data from officials, such as investigators in a fraud case; (4) domestic and foreign wire services, the most common sources for broadcast news departments; (5) competitive news companies, such as newspapers and other broadcast and cablecast facilities. These are the principal sources for anchors and general reporters. Most reporters tend to be specialists. Some are excellent at current events involving urban problems, for instance. These same reporters may be less impressive in identifying and writing stories of human interest.

The key to success in developing your material is selection. Select a speciality or two within the broad range of news events. Fine arts, medicine, space science, education, economics, and politics are just a few specialties. Weather or sports may be a special interest to you. These two areas have maintained high audience appeal, and so they are frequent entities within a news program. Even though a journalist is prepared to cover any story, so to speak, you may want to be identified with an area or two in which you have depth in interest and knowledge. You may find that further refinement of your material is necessary. In sports, you may decide that, although you can do play-by-play for major sports, you are best at only one sport. Or you may realize that you are better at feature sports or *color* background rather than calling the game. Selecting your material is a part of determining where your abilities and interests lie. The material you select tends to define who you are as a television performer.

Furthermore, as you write your material you will enjoy a sense of precision that develops in your writing style. You will find that your material is right for you to deliver and perhaps no one else. In any case, you know you enjoy reading it to your viewers, because you have constructed it so that they will appreciate it, too. Refinements in writing style, just like careful selection of material in the first place, will not happen instantly. They will emerge as you become experienced in television performing.

The task, therefore, is in identifying material that is appropriate for you to present as a performer, and for you to develop skills in gathering the facts to structure a story, newscast, or other information format. Regardless of content, there is similarity in procedure: gathering data, writing it up, assembling the information into proper formats, and delivering it to the public. Obviously then, television performing is not simply looking attractive on the air.

THE BASIC NEWS STORY

The report or *story* is the basic unit of a newscast. The nature of the report is mainly current information with high impact, called *hard* news, or supplementary information that does not require immediate delivery, called feature or *soft* news.

Journalists attempt to answer, as precisely and accurately as possible, *who* the participants are in the event, *what* happened, *why* it happened, *when* it occurred, *where* the action took place, *how* it happened, and, sometimes, *how much* (for example, money, property, time) is involved. Keeping journalism's "five W's and an H" in mind, the reporter focuses on what the public needs to know. As these facts are gathered by the on-camera reporter, a photographer, actually a videotape camera operator, takes those shots that will correlate with the action of the event. Later, in the editing room at the station, the data and the pictures will merge into an audiovisual presentation; both will explain the event to the public in a concise, understandable narrative. That narrative is usually chronological, but occasionally it is topical. If the narrative is chronological, this simply means it has a beginning, a middle, and an end, often presented in sequential time. To capture immediate attention and to create the aura of excitement, a reporter will probably introduce or lead into the story with the current status of the event. Thus, the chronological order may be reversed or altered; that is, the end of the event (or current status) is told first, and then the reporter fills in the background.

Whether you have data already or you must gather it at the scene, you know the questions the public wants answered. The order in which it will be told is determined. The rest of the story consists mainly of verification, such as testimonials from officials and participants, and perceptive photography. Because a guiding principle in television is "show don't tell," reporters are often on-camera for a brief lead-in and wrap-up, providing voice-over narration in-between. A wrap-up usually projects the future, that is, what in the best judgment of the reporter is likely to happen next. In summary, the basic news story consists of the lead, the body explaining journalism's "five W's and an H," and a wrap-up. The procedure for covering a news story is as follows:

1. Get the story concept and assignment from the assignment editor, who links reporters with the day's stories.
2. Gather the facts of the event on location from the participants, while your photographer is taking visuals of the main action, accompanying testimoni als, and background orientation.
3. Organize the facts, tentatively, into a smooth narrative. This is done while driving back to the station.
4. Combine the narrative and the visuals in the editing booth at the station. Try to maintain an on-camera presence for yourself in the story. If the visuals are poor, you may have to present your story during the newscast.

Hard news events happen in the present time, and have a major influence on large numbers of people. Soft news or features add substance to hard news topics, and may be part of an ongoing public discussion. Some feature stories are humorous or anecdotal with little or no impact on public activity. Even so, a news story follows a typical pattern, whether the event is scheduled and anticipated or unscheduled and unanticipated. Here are some examples:

A Scheduled Story: Hard News	A mayor's state of the city speech. You must glean from the speech the essence of what the public wants to know and needs to know to properly participate in city government. In a hard news story told in present time at a scheduled news event, you lead with what the mayor has said that is most important, such as new city programs, new taxes, or new trends, for example:
Story Structure	
Who What Where How Much Attribution	The mayor (who) wants to build a new dam (what) on the north end of Blue Lake (where), if the taxpayers will vote 10 million dollars in bonds (how much). The mayor says (attribution) the dam is essential to water reserves in the decade

Why/Outlook Speculation Signature	ahead (why). Public debate begins next week. The mayor expects to get her way on this issue. John Doe reporting live from City Hall.
An Unscheduled Story: Hard News	For a hard news story reported in the present time at an unscheduled event, the lead is similar: What happened is summarized in one line, for example:
Story Structure Who What/How When Where Attribution Why Speculation Signature	Sixteen-year-old Mary Brown (who) is dead (what). She was shot (how) about three o'clock this afternoon (when) at her home in the 1600 block of E Street (where). Police say (attribution) she was shot by her boyfriend, 18-year-old Bob Jones. They were playing Russian roulette when the gun discharged (why). Jones is in custody. Charges are expected to be filed tomorrow. Maxine Dayton reporting.
A News Feature: Soft News	It may be a multi-part series on hunger, drugs, or other urban problems. It may be humorous inconsequential footnote on a day's events called a *kicker*, and usually placed at the end of a newscast.
Story Structure When/Who Where Attribution Why Close	The alarm rang (when). Bill Bryan (who) walked into the shower (where) half-awake. He turned on the water. As he stepped onto the floor, he felt it move. He looked down. Bryan says he made a yell Tarzan would have been proud of (attribution). 12-foot boa constrictor was lounging on the shower tiles. Bryan's yell was one of joy. You see, Hefty his pet snake disappeared over a year ago (why). But, today, Hefty is home. That's the news. Thank you for watching.
Sports News:	Sports news must begin with hard news, that is, sports scores of both local and national games. This may be all the time allotted to sports. Usually, the sportscaster manages to include one or more feature items. Celebrity acquaintance is a big part of sports, and so you will want to associate with key figures. Typically, a brief soft news interview highlights a game, for example:

Story Structure	
Who What/When Where (Reporter's main question)	The Ace team's top pitcher Abe Aronson (who) was a major reason for the team's victory tonight (what when) at Home Stadium (where). How does it feel to pitch a no-hitter?
	ARONSON
Testimonial	It feels great, Duke. But no one wins a game alone.
	REPORTER
Game Highlight	The highlight for me was when Able Player stole Third in the eighth inning.
	ARONSON
Expert should avoid repetition in word choice	That was a great moment all right. Every player was great tonight. The other team was tough to beat.
	REPORTER
Comment Future Signature	It's a happy locker room for the home team tonight. Ace team goes against the Big Leaguers on Friday. Duke Davis reporting from Home Stadium.

NEWSCASTS

The news content is determined by the news producers' behind-the-scenes staff, who actually run the program. Early in the day the stories that will be developed for airing emerge. This is an ongoing process. Some stories will be mentioned only in anchor narratives, but most reports will have audiovisual support. Short pieces, cerebral items, stories that are just breaking and have no visuals, or reports that are too expensive to cover will be read on-camera by the news anchor.

Newscasts are similar in structure. First there is a billboard (headlines), and then the top story of the hour (often this is a local story or a national story with strong local impact). The value of a news story depends upon its influence on large numbers of people. Natural disasters, wars, economic changes, and life threatening events are potentially stories of heavy impact. They are hard news and exist in present time only. The objective of television news is to tell what is going on *now*. Stories taking place earlier in the day, last night, and yesterday require updating to current status. News, therefore, is highly perishable.

Because television hard news is so brief—about 9 to 12 minutes in every half hour newscast—the top stories are clustered in two 6-minute segments at the beginning of the program. They are followed by features, sports, weather, and a final human interest story, which is usually amusing and read by the anchor, called a *kicker*. Reporters, therefore, tell their stories in 20 to 90 seconds, but 90 seconds is a long story. No one appears on-camera continuously for longer than 20 seconds. Visuals and newsmakers keep the pace of the news fast and maintain an aura of importance.

Obviously, a reporter prepares a tightly written, easy to comprehend package of narration and audio-visuals with a minimum of on-air exposure for herself. The exception to this is the *live* report from the scene, where the reporter stands in front of or near the event, usually at an accident, fire, or public building. The report is partially on videotape, but anchors at the station also ask questions. The questions and answers are prepared largely in advance and coordinated with the journalist at the scene. The news department cannot afford to have its reporter quizzed without having correct answers. The live report creates the impression that the news team is giving the most recent update, and once the station has gone to this expense, it wants to make the venture worthwhile. Little pondering can be done on camera. Smooth ad-libbing and finesse in talking with celebrities, politicians, and sports figures are essential to your prestige, credibility, and personality.

THE BASIC INTERVIEW

Numerous in-depth programs are primarily interviews or a collection of interviews. Celebrities discussing their lives and careers make up much of the content. Some programs concern historical events, philosophical and religious issues, business matters, public concerns. Although the nature of the content may change, the television performer's role remains fairly constant.

The key to a good interview is the same key that generates a good news report—a concept or *angle*, that is, a fresh approach, one that may suggest solutions as well as identify problems. Interview programs shed a good deal of insight, as well as endure much rebuke, for a tendency in some of them to appeal to the seamier side of human life.

Note the typical structure for a television interview. Let us say it runs from 5 to 15 minutes. Half hour magazine programs may use the shorter interview; one hour shows may use the longer one. Every interview must have a reason for its existence. Why do you want to talk to your guest? What do you have to talk about that the public wants to hear? The interviewer begins with a 15- to 20-second opening to acquaint the public with the guest.

Personality Interview

Interviewer's Perspective	INTERVIEWER
Guest's Name/Title Background (even if famous)	Mary Megastar has been a film actress for over 10 years. She has a handsome, successful husband and two bright children. She seems to have everything a woman could want.
Central Focus Lead Question (the public wants answered)	Yet, last June she attempted suicide. Do you think you can talk about it now?
List Questions	1. What caused you to try suicide?
Cause/Effect:Solutions	2. How have you solved your problems?
Current Status	3. Do you think you'll try it again?
Solutions	4. What suggestions do you have for others with problems?

Closing: Name/Title of Guest Repeat Focus	Thank you Mary Megastar for being brave enough to tell us about your problem.
Information Source/ Promotion Signature	If you want more information, Mary Megastar has a new book, *Survivor*. Until next time this is Helen Hostess.

Issue Interview

Interviewer's Perspective	INTERVIEWER
Lead Central Focus Guest's Name/Title Lead Question	There's more to living on the streets with drugs than just saying "no." My guest today is Local Police Officer Dan Danger. Officer Danger, how many elementary schools do you visit each week?
Topic Outline	1. What message do you tell the students? 2. Do they listen?
Current Status	3. Are you optimistic about the war on drugs?
Closing: Name/Title of Guest Repeat Focus Information/Source Promotion	My guest has been local Police Officer Dan Danger. He believes we are winning the war on drugs, and so do I. If you want an officer to appear in your school, dial the information line G-O-O-D G-U-Y. Tomorrow's community focus is on the
Signature	homeless. I'm Lyle Boggs.

An interviewee should not expose himself or herself to a television interview unless the individual is willing to express his or her views and feelings openly. The self-examination can be intense, so don't get into an interview as the guest unless you are sure you can handle it. Many persons are willing to be interviewed after they have written a book on their problem, not only to promote the book but also because they are fully prepared to discuss it. If the topic is personal, anticipate that the questions you do not want to answer will be asked. Naturally, a good interviewer will ask them, because they are the ones the public wants answered. The interviewee must anticipate all the questions that might be asked. Often, the kinds of questions will be defined in advance by the interviewer. As the interviewee you may try to avoid sensitive ones, but it will be difficult. So be prepared.

Lifestyle Interview

Interviewee's Perspective	INTERVIEWER
Background	*My Two Lives* is a new book by Jon Jones. Jon was injured in an auto accident five years ago, leaving him paralyzed from the waist down. He knows
Central Focus	what it is to be able-bodied and to be handicapped. Jon, what are the biggest problems you face?
Topic Outline	1. Transportation 2. Depression

3. Financial problems
4. Human relationships

Jones is attractive and neatly groomed. The questions are short and his answers are short, too. Jones exudes charm, but he seems genuine.

JONES
Transportation is my biggest problem.
I can't drive, and so I have to depend upon others.
That's tough. You aren't free.

INTERVIEWER
So, do you get depressed?

Jones only answers what he knows/feels
He is optimistic

JONES
Yes. I was very depressed right after the accident. But much of that is behind me.

Jones wants to avoid embarrassing questions

INTERVIEWER
Have you adjusted in normal ways?

JONES
I have a girlfriend. That's fortunate.
We just met. We are just getting acquainted.

Interviewer persists

INTERVIEWER
Can you have intimate relations?

Jones anticipates the question

Jones tries to redirect the agenda

JONES
Having sex is complicated for many people, not just the disabled. In my case it's too early to say. My main concern is being self-reliant. Most disabled have difficulty finding work that's satisfying and having transportation.

INTERVIEWER
If you could say one thing to the viewers, what would it be?

Central Focus
Jones has a clear message for the public

JONES
Employ the disabled. They want to work, too. (JONES GIVES A GENEROUS SMILE.)

INTERVIEWER
My inspiration today is Jon Jones.
His wonderful new book, is *My Two Lives*. We'll be back, right after this message.

INFORMATION FORMATS

Common information formats other than the news report are the interview, the demonstration, and the speech. Any performer who gives a speech as basic material is akin to entertaining. For this performer must be intensely charismatic, so much so that he or she can hold an audience without visual supplements to any great extent. Such persons are certain ministers, teachers, and politicians. In an interview, be it one person to another person, or one person to individuals in a group, visually the result is referred to as *talking heads*. No where is the importance of material better illustrated than in long formats involving talking heads (Donahue, Oprah Winfrey, Geraldo, Larry King). Hot topics (often involving sex and violence) and major celebrities (preferably emerging from major projects) keep these programs on the air. Notice that the programs are similar in formula. The host provides a background summary of the topic, the guests are introduced, the guests (three to five) are interviewed, the audience asks questions, an expert is introduced and lends some perspective to the topic, the host briefly summarizes, thanks guests and audience. In an interview, one-on-one, the typical information pattern emerges: (1) the current status and what the public wants most to know, (2) the background details, and (3) the future of the guest or of the issue. Twenty questions covering this structure will prime an informed host, who knows that only five questions may be covered. If so, they will be the top five, and through careful listening on the part of the host, the interview may be more insightful than anything the producers could have prepared in advance. This assumes that the interviewer is well read. The interview has a 20-second lead-in consisting of why the topic or person is worthy of public attention, the person's name and title as it relates to the topic, and a challenging initial question, followed by others the viewers want answered. Most beginners are afraid to ask tough questions, but you should anyway. Those on the talk circuit expect them, and so does the public. If you don't ask them, politely of course, but clearly and directly, you are not working hard enough on behalf of the public. You represent the public, after all. The closing of the interview is the opening in reverse order—restatement of the question, thanks to the guest, repeat of the name and title or relevance to the topic, summary or tag line for the interview, and the host's signature. Promotion for a future program may be included.

THE BASIC DEMONSTRATION

A demonstration is the third unit of information programming along with the report and the interview. The demonstration is particularly valuable as a teaching tool. After all, besides a lecture, how else does one usually learn? Imitating someone's demonstration is a key learning technique. Whether you want to teach a person how to drive a car, win at golf, or bake a cake, a demonstration will probably be the main means of instruction. An increasing number of video cassettes demonstrating everything from how to operate a fork lift truck to how to set a table require performers skilled in the task. The broadcast and cablecast and video cassette markets that use demonstrations for instruction are huge and growing.

Clearly, most television performers cannot complete the task assigned to them within the time frame of the program. Advanced planning is, therefore, critical. First, the demonstration must be thought out step-by-step. Each step must be illustrated sufficiently for a viewer's understanding. In order to bring the demonstration to completion within the time allotted, the demonstration materials must be prepared in steps of progress. In other words each step is partially finished before the actual demonstration is presented on television.

Demonstration: Instruction	Making A Doll
	DEMONSTRATOR
Memorized Open (for a fluent open/ close)	Hi, I'm Doris Teacher, and I'm going to show you mothers out there how to make dolls out of scraps around the house. If your children want to watch, that's okay, too.
Script Outline	I'm going to make this doll in five easy steps:
All items needed are placed on a table and identified	1. Exhibit materials.
One pattern is cut on-camera; others are pre-cut	2. Cut patterns.
One doll is assembled from materials on table; other examples are completed	3. Assemble doll's body.
Doll is dressed; another doll nearby shows the completed look	4. Dress doll.
Several elegant dolls are displayed to show potential of project	5. Decorate doll. This step like the others is only begun.
Closing Memorized	Isn't she pretty? Making dolls is easy, fun and inexpensive. You and your children will enjoy it. If I can do it, mothers, so can you. Bye.

Instructional programs like the demonstrations themselves are carefully structured. A demonstration is a brief how-to-do-it program. The initial expectation of your viewers is what you are going to accomplish and then the easy steps it will take to do it and the attractive results (which should exceed a viewer's imagination). At the outset you must outline what is going to take place within the time frame you are allowed. Whether you are painting a picture, grooming a dog, or orienting new employees to the operation of a company, you outline what you are going to do and how you are going to do it. Then you follow your outline, step-by-step, and finally tell the viewers you have done what you said you would do. Try not to accomplish too much in any given program. Twenty minutes is the maximum for most viewers in an instructional setting. Ordinarily, television watchers are not dedicated viewers, because messages so often interrupt program continuity. Demonstrations are seldom finished in real time. They are prepared in steps of progress, including the end result.

In conclusion, you should have noted that your material is constructed from basic units: the report, the interview, the demonstration, and the speech. These units, when arranged in a certain order, form newscasts, talk shows, magazine programs, and instructional presentations. After deciding what you want to do as a performer, and what you want to share in news and information with your viewers, you are ready to consider ways to reinforce your material through the consistency of your image.

2

The Look

In the business your appearance is referred to as your *look*. This primarily means your face, but it can refer to your entire physical appearance. The reason a person's face is of such concern is because the camera spends so much time on it. A face should be memorable, expressive, and desirable enough to get repeated invitations to return to the viewer's screen. Your face conveys messages, spoken and unspoken, by simply being on a television screen. Your objective is to control your face and the messages it sends, and to, likewise, control your entire body. This chapter suggests ways to define your look, and the next chapter concentrates on improving it.

Ideally, your look is consistent with your material. If you aspire to be a news reporter, then you should have the mature look of credibility. Yet, you want to demonstrate energy, understanding, and empathy for human problems, constructive curiosity, and fairness in that role. Professional television performers who hold major positions are usually easy to recognize because their look is unique; it is different from that of anyone else. Therefore, your objective is to be a unique looking individual who anchors the news or contributes features differently from anyone else. The viewers will decide whether they like you and are willing to see more of your work. If they do not like you, continue to do the work you believe in. Many fine performers have educated the public to appreciate them; of course, this may have taken 20 years. If you are a dedicated performer, you must realize that recognition may take years or may happen overnight. Regardless, you steadily improve your material and your look.

ESTABLISHING YOUR DOMINANT IMAGE

In order for you to perform well, you will have to settle on an individual appearance that you can live with. This may not be as easy as it sounds. Some individuals are quite happy with who they are and how they look. If your appearance is consistent with your conception of your personal and professional life, then you are indeed fortunate. For many others this consistency is not so obvious and needs some adjustments. You are basically seeking a dominant image that the public will admire and remember. Your image may not, alas, express the true depth and dimension of your soul, but your image is what you are for your television appearances.

Perhaps you do not know how you appear to other people. Maybe you do not know what values others see in you. A simple technique for helping to identify what other people admire, or possibly dislike, is to walk into a room of 20 strangers and ask them to write down a few descriptive words of what they see, and what reaction they have, as you stand before the group. Strangers do have an immediate impression. This impression can be helpful in determining your dominant characteristics, those you may want to improve or eliminate. Key words for a journalist might be: intelligent, serious, mature, credible, articulate, and attractive. Descriptions for feature specialists may vary greatly: funny, sporty, clever, unusual, sexy, perceptive, and pleasant.

These few opinions merely suggest whether you are on the right track in projecting the dominant image that is consistent with the television performing you do or aspire to do. Always maintaining your common sense, now consider in detail your dominant image.

THE FACE

Almost everyone wants to look pleasant. According to an old saying, "Beauty is in the eye of the beholder." This means that physical characteristics you prefer may not be so appealing to someone else. What you consider unattractive and highly noticeable, another person may ignore. How many physical characteristics do you remember about your close friends? What you remember is important to you. Try to be objective in considering your facial assets and liabilities and the rest of your body.

ASSESSING THE FACE

The easiest way to assess your face is to have three 8 x 10-inch close-up photographs taken of it: a front or "head" shot, right, and left profile shots. The photographs should not be retouched. Study these pictures. Notice the shapes of the head and face, features, muscle control, complexion, and hair. Be objective, avoid being judgmental.

Remember that television standards, and standards of beauty for the most part, have been dominated by Anglo-American appearance for Caucasian men and women. The classic Caucasian standard has been preserved in European art for centuries. African and Asian standards may vary greatly, but to this time these standards of beauty have not been defined, discussed, and popularized extensively. It may take decades of international television programming for these standards of beauty to be understood and accepted widely. In any case an attempt will be made to recognize some of their characteristics.

Shape

Traditionally, a pleasant face is described as primarily oval, round, square, or heart-shaped. An oval face is considered the most pleasant for women, and an oval face with a rather square jaw bone is favorable for men. The classic face familiar from Greek statues illustrates that the face is divided into almost equal thirds: one-third from the top of the forehead to the eyebrows, one-third from the brow to the tip of the nose, and one-third from the tip of the nose to the tip of the chin. High cheekbones, deep-set eyes, somewhat angular jaw lines, small, well-proportioned noses, and healthy coloring in skin and hair are admired qualities.

The fact is that most people have some but not all of these qualities. Those relatively few persons who have all of these qualities are considered attractive, even beautiful. An equal fact, but much overlooked, is that variations on these dimensions may in the long run be more memorable, and, therefore, more suitable to television. Many Asians have smoother, gentler facial lines and sometimes less dominant bone structure than many Europeans and Indians. These gentle lines contribute to equally lovely facial structure for Japanese, Indonesians, and several African races. Many of these faces are in concert with the physical extremes of slim, short people or very tall, slim people. The point is that on the international scene, the judge is in the eye of the beholder, and there is a greater variety of opinion than ever before.

Features

While human beings have features—eyes, noses, mouths—generally in the same place, the location varies slightly. Notice that the right side of your face is not a duplicate of the left. Your right eye is similar to, but not exactly the same as, your left eye. These differences, however subtle, have given rise to the notion that a person may look better televised from one side than from the other, and this is true. For most performers the eyes are the dominant feature, because the audience seeing the expression in the eyes—often in close-up—interprets the credibility of a message mainly from the eyes. The English language depends on nuances in delivery to convey ideas and feelings, and these vocal nuances are reenforced by the expression in the face, critically in the eyes. Although most people have dark eyes, blue eyes are a popular choice, yet other colors may be more useful in remembering an individual. The main thing is to have eyes that are energetic, that sparkle with life and expression.

The second dominant feature—some believe it is first—is the mouth. Surely, the most important movement a performer should learn to control is the smile. No action signals welcome, human approval, warmth, or friendship so universally as a well motivated smile. A mouth surrounded by well-shaped, rather full lips, that reveal white, straight, even teeth is most pleasing. In some cultures lip size varies from thin lips to very full lips, and teeth may be augmented with decorations of gold and diamonds. Like the eyes, the mouth can be positioned to convey a wide range of meaning. Performers learn to use the mouth for full expression.

The nose is often the most troublesome feature. It is either big, too small, or unevenly shaped. Too much bone or cartilage at the bridge or too much thickness in the cover skin are reasons for the nose to be out of proportion with other features. Yet, in some cases the nose may be the most memorable facial characteristic. A distinguished feature may not be beautiful, but to the public a memorable feature is more important than beauty or conformity, and a wise performer does not forget that.

Cheek, chin, and jaw bones are important to the shape of a face. Men like to have strong, angular, dominant cheek, chin, and jaw bones. Women need not have such strength in these areas, but many of the classic European beauties do indeed have these characteristics. Notice that modern comic book artists draw heroes and heroines with such strong characteristics. Villains, deviants, and monsters have weak facial structure. Weak chins, jaws, and flat cheekbones are accompanied by big noses and misshapen teeth. It follows that if a performer wants to gain a reputation in news and information, facial characteristics should not detract from the messages you are attempting to convey.

Muscle Control

Muscles enable the performer to consciously and perhaps unconsciously control the face. Smiling on cue requires muscle control. How big a smile? How expressive is it? How crooked? How pointed? How motivated? Speaking on cue requires muscle control to shape words and emotional energy to give them meaning. Muscles surrounding the eyes are very important. Eyebrow movement, opening and closing the eyes, and winking or blinking on cue enhance or detract from the expression in the eyes. Quivering or muscle spasms, especially if involuntary, can cause problems for performers.

Aging changes the appearance and the muscular movement in the major facial areas of the eyes and mouth. Aging eyes may provide a softness in the face that is attractive until the muscles begin to sag. In advanced age the facial muscles lose their elasticity and gravity seems to pull the face downward from the forehead, to above the eyes, to a shallowness in the cheeks, to a drooping mouth, to a prominent sagging neck and chin. Depending upon the role, whether you are a young or mature journalist, aging has a proper place in a performer's work. In general, muscle tone that enhances a youthful, energetic look is encouraged for everyone at every stage in life.

Complexion

A healthy complexion consists of taut, supple skin evenly colored. The objective of adding color to your skin is to improve its clarity and energy. Fair-skinned persons can effectively change their skin color. A favorite color for many men and women is a medium tan, because it looks healthy, especially when heightened by a slight, natural blush. Equally attractive is a stark white complexion that is highlighted by a touch of ivory or pink. Deeper skin colors such as blue-black, dark brown, brown, and some tawny hues of Asians require little or no make-up for television. Dark complexions depend upon clever lighting experts to create a proper three-dimensional appearance.

More troublesome than the basic color of one's skin is damage to it. Scars, blemishes of all kinds, poor color, heavy beards, moles, large pores, fine tumors, discoloration, blood vessels, scaling, and various allergies are among the problems some performers have with their skin. As if these problems weren't enough, various cultures treat skin differently. Sometimes youths tattoo their skin, and thus, may permanently rend it marked. For television purposes, tattoos have no benefit. Nevertheless, all of these potential problems should be listed when you assess your facial liabilities. You must know what problems you have to deal with before you can intelligently overcome them.

HAIR

Ample, long, silky, healthy hair that will remain in good condition and retain its color for a lifetime is what most people want. Such a desire is far from the reality, however. Within a year after birth a person gets a full head of hair that remains full until the late teens or mid-twenties. Men—particularly fair-haired men—often experience substantial loss early in life. However, hair loss, which is believed to be hereditary, may occur in anyone. Women, who are fortunate in having less vulnerability to hair loss, may have deterioration and thinning after repeated permanents.

Hair has many characteristics including color, luster, thickness of strands, abundance, and manageability. Because so many races have little or no hair or experience early hair loss, numerous remedies and much research is underway to improve your prospects of having and keeping your hair. A practical solution for performers is hair products—wigs, toupees, falls, and other hairpieces. Women have made substantial use of these products in the past, and men are beginning to. Male celebrities have depended on hairpieces for centuries. Though self-conscious about it, more men in public life find that wearing a hairpiece is desirable, even though it may impinge on their credibility. The point is performers—male or female—must have the strength to decide for themselves whether their hair looks good enough for the role they wish to have in media. If a sportscaster looks better with a toupee, then he should wear one; if he looks better without it and feels comfortable, then he should go in that direction.

Hair coloring, permanents, and adornment for both men and women have come into greater prominence in recent years and are likely to become more so because more men and women want to retain a youthful appearance as long as possible. Nowhere is this more evident than in those who assume leadership roles in media.

ANALYSIS

The key to appearance on television is having or cultivating a unique, and, therefore, memorable look. Visual recognition is the dominant identifiable aspect of television. It is more important than voice, even though that is very important. Your appearance should be unique, clearly different, if it is to be easily remembered by viewers. Being exceptionally beautiful, handsome, ugly, strange, exotic, large, thin, nervous, or energetic can be the basis for a memorable appearance. Your look consists of your face and entire body. Television being a close-up medium concentrates mainly on your face.

After this brief discussion of the optimum characteristics for your face, trace the three photographs, mentioned earlier, to get a sense of your facial structure. Note your assets and liabilities along the lines summarized below:

What is your dominant appearance?
What is the shape of your face?
What is the color of your eyes?
What shape are they?
What is the shape of your mouth?
What is the condition of your teeth?
What is the shape of your nose?
What is the shape of your chin? Jaw? Cheekbones?
Do you have good muscle control in your face?
What color is your complexion?
Describe its quality.
What are the characteristics of your hair?

Virtually every facial characteristic listed here can be altered through make-up, lighting, or medical treatment. First determine your assets, then list your defects. Third, list what needs to be changed, and if corrected, what benefits they will bring to your facial appearance and to your work as a performer.

3

Improving Your Look

Once you have evaluated your appearance, you are in a position to decide what, if anything, you need to do to improve it. Presumably your assets far outweigh your liabilities; but to be effective as a competitor with the best performers in your field, you want to improve whatever deficiencies you have identified. Few, if any, deficiencies are insurmountable. You must find a way to overcome your less attractive qualities by altering them, by reducing their impact, or by utilizing them in an innovative way. Remember such a positive attitude extends to the disabled and elderly; because as years pass, they too will be on television in increasing numbers, and those who have inner faith and energy will do a wonderful job.

IMAGE CONTROL

Keep in mind that television does not reveal the whole truth about a performer, just part of it, Hence, your objective is to show what you want to show and nothing more. This procedure is image control. To control your image, you must be aware of the image you have, and the image you want to have for the work you want to do in news and information. After an analysis of your look, what are your overriding characteristics? Do they constitute the image you prefer? Is this image right for your professional future? Don't be hasty in making this judgment. Let us say, for instance, you are a humorous, chubby fellow audiences adore; however, you would prefer to resemble prince charming. Nevertheless, you accept the image of a likeable comic; and consequently, everything you do and develop for television supports this comic image. This image is right for you, is consistent with your assets, and is admired by your viewers. You, therefore, accept this fine talent for what it is. Soon you find that this asset enables you to sell products and services, to do sports features, to add a flare to weather reports, or to be an effective TV teacher for high school students.

Controlling your dominant image begins by allowing others to help you find that quality that is fundamental to you, and then performing on television gives you the chance to express it. Most likely your image will come from within yourself, and so you must diligently hunt for it. Sometimes you will find your image by defining what you like about other people. What draws you to someone else? Being friendly, open, smiling, and good natured are qualities that form a favorable image for some persons. These characteristics attract admirers. But your dominant image must be within you; it must be your own. You must find it, develop it, and present it on television.

After you find an image you are comfortable with, write material that is consistent with it. Set aside other material you have. Focus your attention on what you can do best. Prepare and rehearse to perfection. Naturally, you hope to work with highly capable directors and crews and to enjoy the expertise of an excellent editor. You must appreciate that there is a certain amount of trial and error in what you do, and there is a great deal of luck and opportunity that must come your way. Even so, determining your dominant image and selecting the best material you can get a hold of will add focus, direction, and momentum to a specific career objective such as news anchor, contributing reporter, feature information specialist, or host of a talk show.

Anyone who considers the political arena must recognize the value of a dominant image that will carry him to office. A winning politician must radiate personal positive qualities. Trustworthiness is right at the top. The politician must convince voters, largely by means of television, that he or she can do the job and be trusted to do the right thing. This image is created, in part, through proper make-up and clothing.

MAKE-UP

Assuming you are healthy and your body is in good condition, you won't need much make-up. The objectives of make-up are to keep the healthy, energetic color you already have and to cover up any unsightly areas, reduce shine, and perspiration. The rule of thumb is to apply make-up that will reach these goals.

For professional performers greater scrutiny is necessary because of your frequent appearances on television. This scrutiny begins with your skin. Normal treatment of your skin is to wash it at least once daily with a mild soap and warm water. Because retaining water in your skin plays such an important part in keeping a youthful look, various beauty plans have been written up that ultimately involve sealing the surface of your skin with a cream that will retain water. Various plans offered by prominent cosmetic companies should be studied. Avoid products that dry or irritate your skin. Common blemishes can be eliminated with over-the-counter products containing benzoyl peroxide, sulfur, resorcinol with sulfur, or salicylic. They are found in a pharmacy. Substantial acne may leave scars and painful inflamed areas that need medical attention. Eruptions may be caused by allergies related to common food, animal hair, dust, and pollen; and even sunlight, grass, and your own perspiration. For many performers skin allergies may be common, and the use of hypo-allergenic products can minimize any risks of irritation to the skin. Most television professionals have their own make-up supplies, but they rely on studio make-up artists to blend the proper effect. Allergies may be encouraged by nervous tension and strain of media work. For a modest sum, a consultation with a dermatologist or an allergist can ward off substantial skin damage.

Television lights tend to drain energy from a performer's face, especially pink hues in fair-skinned persons. To reestablish these tones, apply a base that is either the same color as your natural skin or one or two shades darker. Experiment with several colors before selecting those right for you. A youthful person tends to look darker and redder; an older person looks paler. If you have a natural tan skin, you may only need a foundation that matches it so that you can cover up beards, freckles, or discolored areas. If your skin is black and oily, you will want a matching foundation to dry up the oil, producing a matte finish. Do not highlight your face with a white powder. All performers should buy one to three water based foundations of good quality (one matching, one or two somewhat darker than the normal skin color) that are durable under hot lights, smooth after application, blend with your complexion, easy to apply, and render a nonreflective, matte finish.

A sure sign of a careless performer who has lost the competitive edge is one who neglects the make-up, and thinks the public is willing to see facial imperfections.

In applying make-up for typical use, men wear a waterbase foundation that is evenly distributed over the face and ears and under the jawline and chin. Usually no make-up is applied to the neck, if it is near to or touches clothing. Difficult blemishes get two or three coats of foundation. No make-up is put on the mouth, and much of the foundation is thinned out around naturally colorful areas like the eyes, brows, and lashes. While the beard is thus minimized, the hairline may be darkened by the careful use of an eyebrow pencil on fair-haired men. Men must avoid an artificial, effeminate look from excessive make-up.

Women should identify their most attractive facial assets and accentuate them with cosmetics. Eyes, mouth, and bone structure are emphasized by adding and shading various colors of foundation, blush, and eye shadow. Mascara, false lashes, lipstick, and eyebrow pencil do most of the work. Fashion dictates the added use of a variety of colored eyeshadows and blushers for the cheeks. Natural beauty is highly desirable for television close-ups. Experimentation will enable you to apply the right amount without calling attention to your make-up.

Women should keep in mind that light colors appear to advance, and dark colors seem to recede on television. To make eyes recede, for example, use a dark eye-

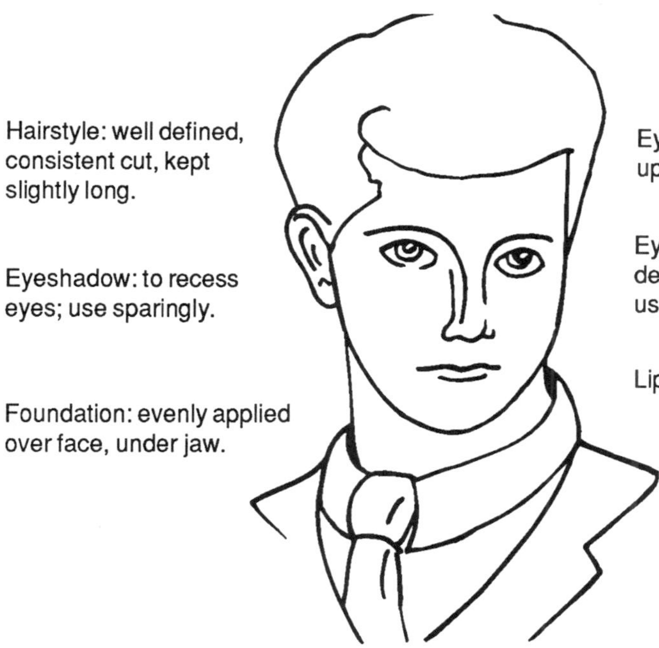

▶ *Figure 1 Make-up for Male.*

▶ ***Figure 2*** *Make-up for Female.*

shadow. To emphasize cheekbones, apply a light foundation above the bone and a darker blusher underneath. Do not apply blusher to the cheeks closer than the mid-point of the eye. In regard to your lips, outline them in a desirable, but reasonable, shape in a somewhat darker shade before adding lipstick. If you want a fuller look, top your main color with a lighter hue or gloss. If in doubt, have your face and hair charted by a well-known professional make-up artist. It is worth the money for the consultation.

A basic make-up kit would include preparations manufactured by an established theatrical supplier. These items are commonly found in a small bag or metal box: waterbase foundations, one matches the skin, two are variations; brown or black eye-brow pencil and sharpener, mascara, various eyeshadows and shades of blushers, lip-sticks, soft brushes in a few different widths, a small mirror, mild soap, washcloth and towel, and tissues or cotton balls.

HAIR

Good skin, natural looking make-up, and clean hair that balances your features round out your facial assets. If your hair is ample, manageable, of good color, and silky in texture, you will probably depend on it greatly to enhance your media image. If you have difficult hair to work with, you may turn to a full or partial hairpiece. Wigs have been available and popular with both sexes for centuries. The key to a good modern hairpiece is how well it blends into the natural hair. A finely crafted hairpiece, made of real hair, and carefully styled is essential to the image of many of the world's leaders and media performers. Performers have them because of insufficient hair, the convenience of multi-styles and quick changes, and the reduction of harsh chemicals on their own hair. Young men who experience premature hair loss should seriously consider a medical or cosmetic remedy.

Hairstyles throughout the world are similar. Although some trends influence youths, particularly for festive occasions, hairstyles for both sexes in media are conservative. Information specialists keep their hair rather short. Few men on the air in news have long hair. Some men try permanents for body and curls, but they eventually settle for straight hair, combed back, and parted. As they get older, many men in media dye their hair. Female journalists are quite conservative in hairstyles, too. A plain hairstyle is easier to maintain. Once she has found a really attractive hair design, she uses it consistently. Typically, a woman's hair has some curl or wave. It is combed straight back and tied or allowed to fall to shoulder length. Frilly, long hair, or exotic, frequently changed styles tend to deny the serious image that most female journalists want to establish. Female journalists tend to be consistent with their hairstyles; this is part of their image. When a more complex hairstyle is required, hairpieces usually meet that demand.

Hair and make-up have a long tradition for enhancing the appearance of those who use them. On the other hand, there is another view that resists the use of these aids because they are supposedly unnatural. Some men are particularly reluctant to use these products. While in office President Ronald Reagan was persistently asked whether he dyed his hair. The point is these aids are used to improve the image of a leader who is seen on television. Busy office-holders sometimes wear make-up most of the day, if they know they are going to appear frequently during a tight schedule. Some leaders achieve a similar effect by maintaining a suntan. Others go to the trouble of wearing a hairpiece so that they will look youthful and vigorous to themselves and to their constituents. Tastefully designed and carefully applied, these aids will, in the long run, be of benefit to a performer in news and information.

CLOTHING

"You are what you wear," according to some commercial advice on clothing. Such sayings may help you choose your personal wardrobe. For most performers in media, conservative attire is best. Medium colors, as converted from a gray-scale, are best; that is, colors that would correlate to the middle range of colors on a scale that runs from black to white. For men, an understated medium colored suit, pastel shirt, contrasting tie, hosiery that matches the trousers, and dark shoes are fine. A further guide dictates that men with black hair tend to favor blue or gray suits, and those men who have blond or brown hair look better in brown suits or other warm colors.

For women, a simple dress with sleeves, plain necklines, a skirt with some fullness of fashionable length, and a mid-range (gray-scale) flattering color is ideal. If a person has light skin, medium to dark clothing tends to set off the face. If your skin is dark, medium colored clothing is preferred. Consultants can often be found in large cities. Many publications are available on the subject.

Mainly for electronic reasons involved with the television scanning, avoid complex patterns, plaids, checks, some stripes, and shiny jewelry. Engineers prefer that performers stay away from shades of blue and green. Blue or green may be dedicated to electronic inserts during the program, such as pictures beside the newscaster. High contrasts such as a white shirt against a very dark complexion may result in a loss of facial definition, as an engineer attempts to compensate for the intense light coming from the white shirt. Do not use your clothing as an advertisement, unless you do so purposefully. Some performers endorse brand name products in commercials, and promote causes. Ordinarily, clothes with scenes, words, or symbols send messages that a television performer in news and information may not intend. The fabrics should not rustle or make noises when you move. Cotton or soft weaves may prove better than corduroy, silk, or taffeta. Avoid buttons and other hard surfaces if they get in the way of body microphones. Avoid bright shoes, mismatched or loud socks, and bare ankles that call attention to themselves. If your objective is to show off the latest fashions, and such a demonstration is acceptable to your message and image, as it is for some talk show hosts, then these clothing extremes are appropriate.

MEDICAL OPTIONS

Improving your look may require orthodontic, optometric, or surgical procedures. There are many medical options available that you may find beneficial. They are designed to enhance your appearance and are generally safe, although no reputable medical practitioner can guarantee results. Once again, you survey your assets, current status, and professional objectives. If, after making a serious study, you decide you have certain problems that need correcting, you seek professional advice in regard to remedies and costs. You owe it to yourself to set whatever fear and apprehensions you have aside, so that you can thoroughly investigate your requirements.

One of the most common concerns is orthodontic. Your teeth should be attractive—white and even—so that you can depend upon a pleasant smile. Proper dental care can make a remarkable transformation in your appearance. Teeth can be straightened, repaired, capped, replaced, and generally made to seem perfect. Depending on the extent of work you require, the cost will vary greatly; but this lifelong investment in your teeth and smile will be worth it.

Another frequent medical consideration has to do with your eyes. As a television performer, you must be able to read a teleprompter or cue cards 10 or more feet away from your position on the set. If your eyes need assistance to do this, you should study the possibility of wearing contact lenses. Many designs and colors are on the market today. Working with optometric professionals, you will be able to improve your sight without interfering with the appearance of your face. Glasses, unfortunately, have some drawbacks for television performers. Glare from lights reflected in the lenses, shadows from rims and bows, and creeping out of position are some of them. Therefore, contact lenses are much preferred. If you cannot wear contact lenses, and some persons cannot, then you may have to resort to glasses, which must be blended into your image. The importance of proper eye care requires in-depth study and is critical to your future.

Further medical options deal with surgical alterations to your face and body. Cosmetic surgery has become popular as improved techniques, better results, and lower costs have made it a more desirable option. Again, thorough knowledge of what can be realistically accomplished and what values cosmetic surgery will provide for you is essential for proper decision-making. The principal reasons for cosmetic surgery are (1) that your image will definitely improve, (2) that the procedures, carried out by a highly competent medical staff, are relatively safe, inexpensive, and available, (3) that this is a lifetime career investment, and (4) that it will increase your self-confidence. Why spend an entire lifetime wanting a smaller nose, larger chin, or whatever, when you can have them while you are young and when they will do you the most good? In later life, why not have a facelift, if that is a benefit to you, your career, and your personal measure of self-worth? The reasons for not having cosmetic surgery are (1) that the procedure will not really make a significant difference in your dominant image, (2) that it may be painful and risky, (3) that your expectation of the results is unrealistic, (4) that your recovery time is slower than your schedule permits, and (5) that it is not affordable. All of these drawbacks must be thoughtfully weighed before you make a final decision. Many of these procedures can be safely completed as outpatient office visits.

In the final analysis, as you evaluate your appearance, you have many wonderful choices in developing your dominant image. Among them are medical options, including orthodontic dentistry to beautify your teeth, optometry to improve your eyesight with discreet contact lenses, and cosmetic surgery to reshape your overall physical appearance. Yet, the majority of television performers will rely upon make-up, hair,

4

The Voice

All human beings produce sound in the same way, by inhaling an air supply and by controlling it when it is exhaled. The air supply is taken into the lungs, then it is released, out a tube in the throat called the trachea. The air stream is shaped or modulated as it passes soft surfaces such as vocal folds, muscles such as the tongue, and hard surfaces such as bone and cartilage.

Recording your voice is very helpful in assessing it. After you have recorded your voice, play it back. You may not like what you hear; you may be highly critical of it. This is a common reaction. Regardless, the main concern is whether you have a voice you can develop. You want a voice consistent with your image and sufficiently flexible to meet your expanding career goals. The main characteristics of a voice are quality, resonance, and phonation. Of equal importance is the way the voice is used; that is, the thought process behind speech called interpretation. Let us, therefore, define and analyze your voice and later discuss the complex task of developing it. As you listen to your voice, carefully note its main attributes. What is the overall sound of your voice? Is this sound consistent with your appearance? Do listeners seem to understand you easily? What aspects of your voice need development? Again, do not let premature judgments be overly critical. Instead, let the following discussion guide your analysis.

VOICE QUALITY

Every sound one produces is a nonsense syllable that takes on meaning only if it can be repeated and consistently recognized. Soothing sounds like vowels usually mean pleasure; jarring sounds like shrieks usually mean displeasure. Language is derived from assigning meaning to vocal sounds. Meaning and the techniques for producing sounds are transferred through imitation. A child imitates the speech of parents, relatives, neighbors, and teachers. So, your speech is the result of imitating those around you. If the models you have imitated all your life have been excellent, then your speech will be excellent, assuming there is no pathological interference. Environments where models have been poorly educated or regional dialects, or both, exist are two prominent deterrents to standard speech.

The most important thing for you to remember is that you can imitate anyone's speech. As you grow older, however, and physical and mental habits are formed, you become less adaptable to changing speech patterns. In order for you to determine whether you have standard speech, you must be able to hear a proper model and compare your speech with it. If you cannot hear a standard sound and tell there is a difference, then you will be unable to change your speech. Some sounds are subtle and require listening to a fine voice model and practicing under a capable coach. A coach can point out your speech deficiencies and provide the correct model. By repeating the words and phrases correctly, and by comparing what you say with what the coach says, you will be able to eventually hear the difference and make a change for the better. This process can be quite slow depending upon how deeply ingrained the errors are in your speech.

RESONANCE

Resonance is generated when the air stream passes the vocal folds or *chords* and vibrates them. This vibration involves the surrounding bones and cartilage. Fixed resonators such as the skull, neck, chest, and rib bones do not change in shape or size without surgery. Adjustable resonators such as the mouth, pharynx or soft palate, and the larynx do change in size, shape, and tension. Vocal folds are left relaxed and open for breathing, producing no vibration, but they are tightened and vibrate during speech-producing voiced sounds such as "ma," or "da." By contrast unvoiced sounds such as "f," "t," or "p" restrict the air steam at the teeth (fricative) or lips (plosive), or both, but no vibration of the vocal mechanism is necessary to make the sound. Relaxation is central to fine resonance. Limbering up your voice through word exercises or singing scales will give you a better assessment of your resonant quality.

The resonant quality generated from the entire vocal mechanism in an individual is called *timbre*. Timbre includes all of the nuances and subtle reenforcement an air stream receives from vibrating bones and cartilage, and from the size, shape, and strength of associated muscles. No two voices are alike, but the sounds they produce are sufficiently similar to enable human beings to understand each other. The reproduction of a single utterance is complex, for each vibration causes other vibrations, some in sympathy with the first vibration, some not in phase with it. Those in phase are said to be in harmony, those that are not are dissonant. Unless a voice is altered surgically, timbre cannot be changed.

Most voices are a pleasant combination of various qualities. Little thought is given to most voices, but in media a distinctive voice may be as important as a unique appearance. A distinctive voice usually comes from its timbre. A primarily mellow voice may skew toward any or all of these qualities: nasality, breathiness, thinness, harshness, or stridency. Simply defined, nasality occurs either by allowing air to escape through the nose or by preventing air from doing so. The latter condition is commonly caused by a head cold that blocks the nasal passages. This is denasality. Breathiness allows more air to escape from the mouth while speaking than is normal. The result is or can be intimate, mellifluous tones that are seductive. A thin voice has insufficient nuances, and thereby, reduces the full, rich qualities associated with a pleasant voice. The quality is common in those who have pitched their voices too high for many years, and in effect have only developed the high range during maturation. A harsh voice is caused by the lack of nuances from primary and secondary vibrations in an entire pitch range, resulting in a rough, irritating and genuinely unpleasant quality. A similar strident voice is caused from strain and tension. The resonators are or consistently have been tense. A hoarse voice is unusual and can be quite useful. This low, foghorn-like register can be an asset, if it is natural to a child or woman in whom one would not expect it. A hoarse voice may be the result of shouting, laryngitis, polyps on the vocal chords, and certain diseases.

PHONATION

Phonation is the precise production of a phoneme or sound. Phonation may be thought of in three ways: articulation, pronunciation, and communication. If a performer is to have facility in any language, he or she must learn its sounds precisely. The objective is to know the standard use of the language so well that listeners cannot tell where you are from or what your racial or ethnic origin is. Sophisticated leaders in news and information who depend on media quickly reach this objective, for, in most cases, regional, ethnic, and racial speech patterns can be eliminated or at least minimized to the extent that they are imperceptible to most listeners. News personnel train their voices so that they have universal acceptance throughout the country. Information personalities, on the other hand, may retain some accent to preserve a specific image. Those performers who elect not to aim for standard speech usually believe that their peculiar speech pattern is more consistent and desirable for their personalities, and that they serve a segment of the general public. It has been said that President Lyndon B. Johnson purposefully spoke with a distinct Texas accent when in the state, but spoke general American when addressing a national audience. Professional news persons, information personalities, and politicians are only three groups that serve as role models for adults and children.

Articulation is vital to phonation. Articulation is a performer's ability to produce sounds clearly. Whatever movement of articulators is necessary for the production of a sound seems like a reasonable thing to do. For children it is. For older children and adults self-conscious habits sometimes limit precise speech. Laziness that restricts the precision with which the tongue, teeth, and lips respond to each other reduces the accuracy of your speech. The result is that you will be difficult to hear and understand, especially if there is competition from other speakers or other sources of noise in the room. Impediments to good articulation may range from inactive use of articulators resulting in mumbling to overactive use of them resulting in mouthing words, in which instance the lips are moved excessively. Conscious effort by the performer can eliminate or greatly improve these problems.

How words are formed rests largely with the exposure a child had to family, friends, teachers, and community leaders. Local and regional dialect is bound to emerge as the child imitates the adult speech patterns. If the adults have little or no dialect, then the child will have none. In a nation as diverse as the United States, dialects, accents, and variations in pronunciation are likely.

If articulation refers to the sounds producing parts of words, pronunciation is the enunciation of the entire word. Aside from certain regional dialects, the principal conflict in the United States has been between a preference in long "a" or short "a" sounds in words such as "tomahto—tomato" or long "e" or short "e" as in "been" or "ben." Media performers tend to say "ither" rather than "either." The longer "a" and "e" sounds probably came from theater speech which emphasized the necessity for clarity when the voice was projected into huge auditoriums before the days of microphones. Some people associate these sounds with culture, higher levels of education, and sophistication. So there is some conflict, though modest, in standard speech pronunciation.

Professional commercial announcers, newscasters, politicians and others who presume to represent great segments of the population and who wish to communicate with all of their viewers use the most generally accepted form of the language. In the United States the standard, many authorities agree, is commonly found in the Middle West and California. Pronunciation guides published by networks and wire services help to acquaint performers with national standards of pronunciation. Regional dialects, racial dialects, and international languages influence a performer's speech. Local names, phrases, and other references may be different from the standard but are essential for effective communication. Current trends in media suggest that dialects enrich programs, so there is less effort to prohibit those with dialects from newscasts and other information programs. The standard of correctness in speech has come under more scrutiny, as more ethnic and racial minorities have found roles on these programs. Whereas in the past southern, midwestern, and eastern speech patterns have been dominant, the influx of Spanish-speaking people and Asians portend additional major influences on pronunciation. Even so, the best approach for the novice or young professional in media is to learn standard general American speech, and depend on widely recognized pronunciation for all English words, names, and phrases. Foreign pronunciation may be either Anglicized or given its correct foreign pronunciation. A media performer who has fluency in more than one language has an obvious benefit.

Common errors in pronunciation are heard frequently on media, more so in information programs than in entertainment shows, because much of information is adlibbed and frequently entertainers are better trained in the language. Causes of errors arise primarily because the performer did not learn the word correctly in the first place. Are you guilty of any of these errors?

Sound substitutions: git for get, sinator for senator, min for men, jist for just, becuz for because, pressperation for perspiration, Illinoise for Illinois, pitcher for picture, whur for where, witch for which, wear for where, what for watt, madder for matter, baddle for battle, pilla for pillow, vury for very, coworse for coerse, liberry for library, groshers for grocers, intertainmint for entertainment, mintal for mental.

Sound omissions: han for hang, son for song, han for hand, slep for slept, goin for going, trus for trust, whi for white, chile for child, spect for expect, fiel for field, an for and, playin for playing, memry for memory, hunred for hundred, lil for little, s'pose for suppose, mir for mirror, tempeture for temperature, genlmen for gentlemen, hep for help, zackly for exactly, las for last.

Sound additions: Acrost for across, cain't for can't, pome for poem, athaletics for athletics, meilk for milk, mis chee vi ous for mis chie vous, palm for palm (no "l"), placeda for placed.

Slurred additions: desks, wasps, clasps, discs, doncha for don't you, valyabul for valuable, wanna for want to, harya for how are you, howjado for how do you do.

Misplaced accents: dé bate for de baté, com par´a ble for com´par a ble, the até er for thé a ter, in creasé (verb) but in´crease (noun), re fill´ (verb) but ré fill (noun), a li´ as for á li as, dís close for dis closé, in fá mous for in´famous.

INTERPRETATION

Changes in the voice occur through mental demands placed on it. This is interpretation. Interpretation is based on logical and emotional responses generated in the mind of a performer when a passage is recited or read. Every utterance is interpreted in some way, whether colorfully or flatly. For instance, the sound "o" can be interpreted in various ways with different meanings: "o" means understanding, "o" means surprise, "o" means dejection, "o" means emotional satisfaction, "o" means the last gasp of a fatal blow, "o" means "I'm thinking." Interpretation of this vowel varies in accordance with inflection or the willful or subliminal raising or lowering of pitch and volume in response to the text. Interpretation depends upon several variables including pitch, range, volume, timing, and style. Interpretation is a subtle and complex art of carefully selecting those elements that will communicate impact on viewers. The more you know of life, the better educated you are, and the more experience you have, the greater ability you will have in affecting those nuances that stimulate response. Performers bring new insight to audiences largely through interpretation.

PITCH

Pitch is determined by the frequency of the wave you utter when you speak. The degree of pleasantness with which that wave is perceived by listeners is its timbre. The objective of speech lessons and electronic equipment is to change or modulate the frequency so that its perception is optimal. Every voice is capable of producing a range of frequencies from high to low, some voices having far greater ability than others. Ideally, these sounds are produced without strain or irritation to the throat. Almost everyone finds a comfortable speaking range naturally; however, because pitch can be influenced through imitation, a voice can be the victim of abnormal placement. For instance, if a young girl imitates her mother's high pitched voice, even though it is unnatural to her, while she is growing up, she may eventually develop a high, thin, voice. This is caused by abnormal placement. A boy has a chance to redefine his voice as he matures, and though a girl's voice may change less noticeably, it can be improved through training. To check your pitch and range, go to a piano and strike middle-C, then imitate the notes that are higher and lower. The extent to which you can imitate these notes with clear sound of good quality constitutes your range. Your singing range correlates with your speaking range. You will note that a range exists beyond the highest note you can comfortably produce. This artificial range of high-pitched sounds is called falsetto, and may also have value in your work as a performer.

Perfect pitch is relatively rare, for your pitch may be influenced by a grating or fraying that dissipates the sound, especially when your voice is tired. A relaxed voice in a rested body of high energy will enable you to produce the best normal sounds. Pitch can be enhanced when it is reenforced under favorable acoustic conditions, such as speaking or singing in a shower. In this instance, the frequencies are reenforced by bouncing off from the walls, adding richness to the basic sound. While you cannot go around with your head in a shower, you can use electronic equipment, such as a proper microphone, to enhance your voice. Interpretation requires a substantial collection of high quality frequencies that the voice can produce on demand.

RANGE

Range is the extent of a performer's ability. When the term is applied to the components of interpretation, you must consider pitch from the lowest sound you can produce through falsetto sounds; volume from barely audible to deafening yells at pep rallies; timing, or the precision with which material is delivered as some words, phrases, and passages are slowed or speeded up; variety or the flare for the unusual or dramatic in delivery; and style, or the way performers present themselves in intelligent, convincing, and credible information programs. The sum of these variations, which are merely hinted at here, constitute a performer's range. The greater the range, the more variations the voice has, and the more useful it is.

VOLUME

Increases or decreases in the loudness of your voice give it force, emphasis, and sufficient audibility so that you can be heard and understood. Generally, performers use only enough volume to do the job, because they learn to conserve their abilities without reducing the quality of the performance. A spokesperson for a commercial may whisper for an intimate, romantic effect; an evangelist may shout for emphasis. Both performers graduate volume in accordance with the moment of expression in the commercial or speech. The objective is to have the volume sufficiently soft so that it can be heard comfortably without straining at the low end, and loud enough so that it is stirring at the high end. For live appearances (often taped for later replay) high volume level seems unlimited; for media automatic equipment, known as limiters, keep volume within prescribed ranges beyond which overmodulation or distortion would occur. If a politician wishes to stress a point, that person typically uses volume. Volume is the most common way to tell an audience to pay special attention. A performer uses volume as an interpretative tool for many reasons: to sell products, to create explosive excitement, to draw attention. Volume can harm vocal chords. In some forms of cheerleading, for example, the vocal folds bang together savagely until a tissue buildup eventually causes a callous to form, frequently causing hoarseness.

TIME AND RATE

Time and rate can be confusing. Time refers to how long it takes a performer to deliver a syllable, word, or sentence. For aesthetic purposes, a performer senses precisely when to utter a sound, such as on the beat, after a laugh, before the action, and precisely when to pause, for pausing can add emphasis to a message. Mechanically speaking, a person delivers lines to the second so that the commercial, newscast, or documentary will end appropriately. This may mean speeding up or slowing down the rate or pace at which the words are delivered. On an average there are about 120 to 140 words spoken in a minute of copy. Many performers read slightly faster or slower, but all must finish on time or "on the nose." Consider how rapidly you read copy, maintaining accuracy and clarity in the delivery. Have you developed a feeling for the time it takes to read several lines of material without looking at your watch? You should develop an innate sense of timing that enables you to deliver material to the second without looking at your watch. After all, television is timed to the second. Train yourself to read with timed precision.

VARIETY

Variety is the use of the voice in an unexpected way: a change of pitch (raising pitch at the end of a sentence instead of lowering it), a change of time (slowly reading a line that ordinarily would be read rapidly), a shift in volume (emphasizing phrases through loudness), or a change of style (giving an audience the unexpected such as stumbling instead of the usual slick delivery). Variety draws upon all of the tones available to an interpreter. It can be fun, because the clever performer brings new excitement to material. Variety should be motivated and not just mechanical or arbitrary. When a coach asks for variety in a reading, a novice typically responds, "Do you want it louder?" The coach answers, "No, say it differently." The performer must use imagination to express the line in an original way. Can you read copy differently? Do you like to bring new meaning to ideas and, perhaps, old copy through the inflection in your voice?

STYLE

If you were to ask each member of your family to recite "Mary had a little lamb," each one would say the verse differently. One might be enthusiastic, one shy, one bored; but everyone would read it differently—that's style. Style is the unique way a performer presents material. Perhaps a performer's dominant style may be described as innocent, friendly, aloof, crisp, relaxed, artificial, neurotic, or bizarre. These terms are attempts at differentiating one performer from another. By means of vocal expression performers define who they are when they appear on media. Typically, whether the material is memorized or read, it begins at a high energy level, tending to burst on the scene (without being out of context), progressing at a somewhat rapid rate, rising to a vocal crescendo, tagging the final line, pausing, then cuing the next segment of the program (or a commercial). The performer's final pose is then retained until he or she is positive the next segment is on the air. You must not break your gaze, pose or position, or add any gestures (sniffing) or remarks that are unexpected or not in the script.

Different personalities are bound to develop an individual manner of presenting their material. Some read the news as though it were part of a long narrative with moments of optimism and despair; some read copy with humor and warmth; some read it in a slightly critical or scolding tone; some read with an intelligent, reflective, and philosophical flare; some read in a chatty, flighty, and sensational manner; others are conversational and sentimental; and still others have a knack for the grave, dramatic, and urgent.

Styles vary noticeably in weather reporting. Some read the weather news as a simple fact, some tell it in a comic vein with gags, some are folksy and include information about recreational activities and meetings, and some go in for flight information and meteorological detail. Why is one weathercaster preferred over another? The information is basically the same, so the answer is in the person's way of reporting, that is, the style. A few sportscasters have magnified their own characteristics to become personalities covering the sports scene. A dapper dresser spins complex yarns about selected events with the finesse of a connoisseur of fine wines. The *color* or feature specialist can contribute perceptive observations and analysis. These performers are usually good story tellers and unusual personalities in their own right, whereas the person who calls play-by-play frequently does so in an unadorned fashion.

VOICE ANALYSIS

Try to describe your voice accurately using the guidelines that follow. Do not be overly critical. Your views will change as your voice improves. Ask yourself:

What is the dominant quality of your voice? Is it pleasant? Unpleasant? Uncertain?

Is your voice consistent with your media appearance?

Does your voice tend to be nasal? Denasal? Breathy? Thin? Harsh? Strident? Hoarse?

Do you have trouble speaking clearly? Can others understand you easily?

Is the problem articulation or the way you form sounds? Or is the problem pronunciation, or due to inaccuracies in your understanding of words and grammar?

Have you considered the many ways in which you can bring new energy, life, and excitement to reading and reciting old material? Have you tried to vary the pitch, volume, time, or style in preparing copy?

Now look in a mirror as you recite lines from a narration or a commercial or a newscast. What effect does the combination of voice and appearance have? Is it the dominant image and sound you want?

5

Controlling Your Voice

Voices are very sensitive to a performer's entire physical condition. In the main, a voice depends on vigorous, internal energy, generated by good health, a desire to perform, a certain amount of personal daring or abandon while performing, self-confidence, and a joy derived from performing for others. Relaxation is the key to a full, rich voice. Often relaxation is not possible because of tension or fatigue. Relaxation is largely a state of mind. If a performer believes that the audience is friendly and wants to see the program, and if the performer has some experience, knows what to expect, and likes what he or she is doing, the performer's voice will be relaxed and reflect it. Some performers are much better at being relaxed than others; they can set aside possible consequences of a bad performance by mustering enormous amounts of self-confidence. Many other performers worry about whether they will be successful. Worry causes tension in the muscles of the entire body, particularly the throat. Tight muscles reduce the vibrations in the cartilage and bones, limiting the richness of tone a performer emits. A tense voice is often harsh, strident, may salivate more, dry up completely resulting in coughing, have less breath supply, may quaver, squeak, or shift pitch unexpectedly. Fatigue is the other extreme. The muscles simply do not function precisely, so the words are poorly articulated. Some voices tend to shift to a higher pitch and are thin when tired.

Techniques for relaxation before performing:

1. Take in a large amount of air. Hold it as long as you can. Slowly let it out. The result may be a settling effect.
2. Push the tips of your fingers on one hand against those of the other hand. This may have a stabilizing effect.
3. Move head, then arms, then upper torso at the waist in a circular fashion. Such limbering up is relaxing.
4. Run in place. This may use up and redirect excess energy.
5. Think of relaxing in a quiet place with soft music playing.
6. Stretch your arms toward the ceiling as you lift up on your toes on a one-two count.
7. Jump up and down like a rag doll to relax muscles.
8. Repeat these exercises, keeping in mind that you are not competing with anyone. You are relaxing your body and mind.

BREATH CONTROL

The objective in controlling your air supply is to inhale a great amount of air and release it at will. You should be able to speak effortlessly for a long time, if the air is properly controlled. The performer, standing erect, elongates the torso so that air can fill the body fully and easily. If you place your hands at the waist with fingers almost touching in front and with thumbs to the rear, and breathe deeply, you can feel your fingers being pulled apart as your body is inflated. A performer must learn to control inhalation, retention, and exhalation of breath. The principal muscle controlling these activities is the diaphragm, located between the chest and the abdomen. An erect torso can take in the same amount of air whether standing, sitting or lying, but the air supply is more difficult to control in the latter positions.

The amount of air you inhale depends upon how much you need. If you fill the uppermost part of your body, that is, to roughly the upper trunk, the clavicle or collarbone, you would be able to pant, for example. More commonly, air is taken into the upper lungs for normal breathing and speaking; this is the thoracic or chest cavity. Lungs are quite expandable, and the lower region is used when an effort is made to relax the diaphragm sufficiently to fill the lower lungs. The diaphragm acting like a bellows helps to create the partial vacuum, which draws air into the body and pushes the air out again on command, with assistance from the intercostal muscles and the rib cage. Inhaling should be as quick, quiet, and complete as possible; it should be reserved for ends of paragraphs, lines, or phrases. Only amateurs gulp air within words or phrases or exhaust themselves in their delivery.

Breathing is not merely a matter of getting air into the body. It is, more importantly, getting air out. On expulsion, the air supply, shaped as a stream, is released in miserly fashion so that the performer inhales less frequently, and uses the air more efficiently. Here are some techniques for improving breathing:

1. Practice enunciating vowels within a few inches of a lighted candle without having the flame flicker. If the flame flickers, you are using too much breath.
2. Read aloud long monologues from Shakespeare allowing a minimum of inhalations. Shakespearean monologues help articulation, too.
3. Sustain a note or sound for as long as you can. Time yourself, so that you can sustain sounds longer.
4. Read aloud the old children's poem, "The House That Jack Built." Each verse gets progressively longer, and therefore, is a fine exercise in increasing your breath control.

These exercises, when practiced daily, will enable you to limber up your voice.

INTERPRETATIVE CONTROL

The degree of vigor, brightness, energy, nuance, and meaning that an utterance has relies on the mental stimulus that regulates the various parts of the physique. The reason viewers like to watch even a person of limited physical capability is because of the mental control and energy that the person exercises in completing the task. Often the elderly and physically handicapped are fascinating performers because their mental powers exceed their disabilities to the point that, in some cases, the handicaps are forgotten.

So a performer continuously hears: "Think what you are saying while you are saying it." This mental activity contributes so much that is involuntary that without it, a performer's message may be totally ignored or dismissed as boring. Mental control is largely the realm of interpretation, which is conveying what the piece means, at least what it means to the performer. A delightful exercise in conveying meaning through the mind is to have a conversation limited to a single syllable such as "ah" or "mm." Your imagination can quickly conjure up ten or twenty meanings with discernible inflections. The English language is very dependent on mental inflection and interpretation. A news or information specialist must be well aware of the importance of this mental dimension to delivery and learn to use it effectively. These techniques for interpretation may be helpful:

1. Polarize your audience by composing yourself, frequently pausing slightly before you speak.
2. Your first words or action, known as the *attack*, should be strong in volume and decisive, though not necessarily loud or overly intense.
3. Learn to deemphasize or "throw away" words of no consequence such as "a" and "the." Both should be pronounced with a short vowel rather than a long one most of the time.
4. Collect some excellent scripts in the principal categories of commercials, news, and documentaries. Practice reading them aloud, changing the emphasis to add better understanding and variety.
5. Deliver the last line in a sequence, story, or scene with slightly greater volume and intensity. This is the "tag" line. It tells viewers that sequence is ended.

LANGUAGE CONTROL

Language control concerns your ability to articulate, enunciate, and pronounce words properly. It also has to do with phrasing and grouping words to facilitate a listener's understanding. An intensive study of language is essential for one who aspires to a leadership position in media. Those who are not proficient in language are caught apologizing for their errors on the air. Such errors are heard rarely at the network level, but all too frequently at the local station level. How many times have you heard newscasters stumble over copy? Perhaps they do not speak clearly enough to be understood. A word is slurred, the ending is cut off, the "i's" sound like "a's" and the "t's" like "d's." Perhaps the reader simply does not know how to pronounce the words correctly. Names of persons and locations are particularly troublesome. Perhaps the announcer is poorly trained in English grammar. For example, one sportscaster in a major television market said he felt "badly" (instead of bad) three times while reading one paragraph on the air.

How do you know whether you are having difficulties with language control? Try these techniques for articulation, enunciation, and pronunciation.

1. Record five minutes of television copy. Two days later, after you have forgotten it, play it back with the recorder about 20 feet away. Write down every single word

or syllable that is not absolutely clear and analyze why you could not understand it. Ask a friend to do the same thing: write down every word that is unintelligible. The list that results will suggest how proficient you are in reading aloud. A good announcer is able to read rapidly when called upon to do so without making an error. These performers can deliver messages flawlessly under the duress of extreme conditions.

2. Make a list of common English words that contain vowels, consonant, and diphthongs in initial, middle, and final positions. Record your list. Compare your speech with that of local media leaders.

3. Improve your articulation by reciting these classic lines:

How much wood would a woodchuck chuck in the woods if a woodchuck could chuck wood? If you were a woodchuck and would chuck wood, would you chuck wood like a woodchuck or would you chuck wood differently? No doubt, some woodchucks chuck more wood than other woodchucks.

Peter Piper picked a peck of pickled peppers. How many peppers did Peter Piper Pick, if Peter Piper picked a peck of pickled peppers?

A big black bug bit a big black bear.

4. Limber up the muscles around the lips, tongue, jaw, and soft palate by stretching your lips, whistling, blowing, extending your tongue, and moving your jaw in various directions. These exercises should help improve your articulation.

5. Get a good recently published dictionary. Whenever you have a word that gives you trouble, look it up. Begin with "athlete" and "theater," two commonly mispronounced words. Have a recognized speaker, whose voice you admire, record your list of troublesome words. They will serve as a model for you to imitate. Compare your pronunciation with the model. Mispronunciation is frequently caused by incorrect spelling or faulty hearing. Carefully break down the word into syllables. Try to reconstruct the word. Listen to the model again. Repeat the correct form until you have it firmly in your mind. Don't be impatient. This may take a long time.

AMPLIFICATION

Amplification is magnification of the voice so that the audience can hear better. It can be done electronically or through voice projection. Projecting your voice is not merely a matter of shouting or increasing the volume, although this sometimes accomplishes the goal. Projecting requires that the performer *think* where the voice should reach, and through thinking sends it there. This technique is effective in an auditorium wherein the performer can project or think the message to the back of the house or balcony, and can elongate certain words as one might in theater so that they will carry the distance.

Usually projection is not necessary beyond a minimum utterance, for the tiniest murmur can be amplified with the right equipment. For all practical purposes, however, you should project your voice about 7 feet or to the distance of a studio camera. Such projection will give your voice energy. Amplification enables a vocal range to be used from the lowest levels to the loudest, and these levels can be electronically reinforced and extended, so that the performer's voice will sound even better than the performer could possibly do in live performance. In recent years this has forced performers, musicians especially, to transport their entire electronic package on tour in order to preserve what the public identifies as "their sound," which in fact is a sound produced in a studio, and sometimes cannot be reproduced live on stage without being supplemented, at least in part, by recordings. For example, some well-known vocalists and choral groups are accompanied by prerecorded background music for live and local television appearances. Often when a performer sings alone on a large stage or in a TV studio, no microphone is in sight. He is prerecorded. Dancers, breathless after a strenuous routine, mouth words to prerecorded lyrics. The electronic manipulation of sound is a study in itself, and a performer should get involved in it.

UTILIZATION OF MICROPHONES

At this point it may be beneficial to mention how microphones are used in news and information programming. Microphones as equipment are discussed further in a chapter on the studio. The objective in using a microphone is to provide a clear, noiseless reproduction of the voice and related sounds. In fact, audio for news and information ranges from high-quality to virtually inaudible moments in a soundtrack that has been poorly recorded. Hazardous circumstances, such as criminal investigations, and amateur recordings, such as at the scene of unanticipated natural disasters, may have poor audio quality. Events that demand actual details, even though sophisticated equipment and reporters have not reached the scene (war zone, earthquake), may be fed live to news programs via radio signals, although the only available video is a slide. These circumstances are comparatively infrequent with today's high speed technology and transportation.

At the same time, studio controlled reporting has declined, and remote uncontrolled location reporting has increased. Therefore, the difficult problems of location reporting make the sound quality less than desirable. Some talk shows have infused location staging outside in amphitheaters and docks, where sound is jeopardized. Then too, some performers are wearing wireless microphones which enable them to walk around: but because the sound is transmitted from the microphone, it may fade from time to time.

Dominantly, news and information performers sit or stand. They do not walk very far. Assuming that the microphone is to be hidden or minimized in this limited staging, performers learn to work with whatever is provided. The performer's relationship with a microphone is primarily one of placement. Typically, microphones are of sufficient frequency range to capture and to reproduce the full range of human sounds. An engineer will control the performer's volume with the understanding that the performer will articulate clearly, and project adequately no matter where the microphone is located.

Most situations require animation but little real movement. If you are sitting behind a desk, you may use a desk microphone; that is, a mike mounted to a short stand placed on the desk slightly to one side. In this case your main concern is to know the shape of the beam of the mike and to be certain you speak into it. Occasionally the mike is suspended overhead and at a 45-degree angle in front of you on a long pole-looking device called a boom. Your problems are the same relative to the mike. The advantage is that you keep your head up with a boom-mounted mike and your head tends to go down with the lower position of the desk mike. The microphone of choice these days is the tiny lapel-mounted lavaliere. The lavaliere is attached to a cable. You must keep this incumberance in mind, because pulling on it in any way may reposition it or create unwanted noise or both. Sometimes performers even forget to put it on, while off-camera someone is frantically motioning to them that they cannot be heard. An additional disadvantage with the lapel microphone is that it never really looks attractive pinned to a dress or even on a suit or tie. The lapel cable is highly visible, and you must learn to discreetly manipulate it. Do not let the cable pull on your clothes or trail behind like a long rat's tail. Plan in advance so that you are not wildly throwing it around, so that you can move from the anchor desk to the weather board for instance, if you are doing weather.

Rarely is a microphone held on a lanyard or necklace-like cord. This instrument is even more obvious to the viewer. A headset-microphone combination is occasionally used, if you are not going to be seen a great deal while you are in the press box. If you are standing, the microphone may be in front of you on a shaft anchored to the floor. If so, be careful not to kick the shaft. It is probably metal and will amplify a resounding thud. The floor microphone relationship is the same as other static microphones mentioned above.

If you are a participant in a panel discussion, do not move the microphone(s) that may be present. Let the engineer do it. These microphones are usually omnidirectional, so that everyone can be properly heard. In some interviews you may have a handheld microphone. Never let your guest take the microphone away from you. If you do, you have temporarily lost control of your sound source and your program.

Movement creates several problems for the performer and the engineer. The solution may be a microphone with a long cable, a wireless microphone, or overhead boom microphone. Whatever it is, you must be prepared to relate to it effectively. Mentally you will find yourself projecting your voice about 10 feet; this mind-set will give clarity and energy to your voice in most situations. Again, rarely, you will have to use multi-microphones by more or less going from one type to another during the progress of the program. With this staging your basic problems do not change, they just come faster; and you learn to move from one microphone to another without the viewers noticing.

You will find that about as much of your work is recorded as is live or live-on-tape. Recording allows for precise editing and the increasingly high standard of visual and sound perfection. Depending on your role in news and information, you will be narrating voice-over continuity and synchronizing your words to your image in the studio. This is referred to as post synchronization. Your main problem here is to maintain the same vitality and the same mental process you had during the original presentation. Remember to think what you are saying as you say it. Your objective is to give "live" presence without your viewers being aware of the equipment necessary to do it.

6

Movement

In visual media perhaps no movement is more important than a smile, assuming the smile is appropriate to the occasion. Smiling varies from a minimal grin to fully parted lips exposing most of the mouth. Universally a smile conveys warmth, friendliness, and confidence. A smile is a sign to the viewer that you and the program are under control, even under difficult circumstances. For some people a smile must be practiced until it appears natural and genuinely expressive. A pleasant smile cannot always be generated; and so as a performer, you should develop a smile that appears natural and motivated. To do this, (1) look in a mirror to establish a smile you like by observing different poses, (2) practice the position of the smile until it can be duplicated on command, and (3) display the smile whenever it is called for, knowing that it is attractive and acceptable, even though it may not be motivated because of the stresses which you do not wish to reveal to the viewers.

Movement is more important and complex than most performers imagine. Setting the body in motion enables you to convey perceived and occult messages obviously and subtly. For example, if in a demonstration you pick up an art object, the arm movement is clear. As you grasp it, you show the degree to which you relish the object. You may touch it reverently or casually, depending on whether you think of the item as priceless or a prop. Close observation through the television camera lens magnifies your movements. Are they stable, hesitant, insecure, or eager? No matter how you move, it has meaning. Similarly, in an interview you convey warmth, friendliness, intimacy, and concern, if you are in close proximity to your audience, relate to your director, and seem relaxed while speaking with your guest.

Movement begins and ends in a position of firm posture—standing, sitting, kneeling, or lying. The figure is in a posture that seems natural; yet it is not. Unless your image requires it, you cannot be so relaxed that you slump, droop your head, cave in your chest, or stick out your stomach. Instead, you must stand and sit in a state of heightened naturalness with muscles only slightly at ease. A taut military stance is too rigid for most television situations; a straight body held in somewhat easy posture is preferable. Relaxation is achieved when you have control over your body and its movement. If you were to sit in a natural and relaxed pose on camera like you sit around the house, you would probably look like a lump of clay. Such relaxation may be appropriate for a conversation at home, but it makes a poor on-camera appearance.

Surprisingly, performers move relatively little on camera, in part because the camera moves too. The principal action takes place in the center of the screen, and that action must be instantly understood by viewers. Subtle movement that is important to the presentation is often shown in close-up, which means a performer has far less distance to go than on stage. In any case, movement on camera is or always should be motivated. If there is no reason for moving, then do not move.

The best way to check your movement is to make a short videotape. The videotape should show you sitting, getting up, walking to the center of a lighted area, saying a few words as if you were hosting a program, using a few gestures, and returning to your chair. Another common situation is sitting at a desk reading a newscast or conducting an interview. Taping these activities will make it possible for you to analyze your movements. Watch the videotape carefully, noticing your principal movements. Do you have energetic, natural movements? Are you conveying subtle or unnecessary messages by means of movement? You may need an opinion from a coach or other objective outsider, because you may not be aware of how you look to others.

Research indicates that nonverbal cues may be more than four times as effective as language cues in transmitting attitudes. One study claims that body language can take up more than 90% of one person's message to another. Body language relies on appearance, nonverbal sounds, postures, movement, length of message, timing, space relationships, and touching. The study of all body movement of communicative value is called *kinetics*. Research further indicates that movement has no meaning in isolation, that extra or random movement blurs messages, and that movement can contradict a verbal message. A discussion of movement requires observation of the moving figure, motivation for movement, characteristics of movement, elimination of involuntary movement, and consideration of the tenants of good mental, emotional, and physical health that support movement.

THE MOVING FIGURE

Every movement has meaning. There are two kinds of movements: those which reinforce the message and those which do not. Those movements the performer needs give animation, interpretation, emphasis, clarity, and style to the performance. Those a person does not need blur the message the performer is attempting to convey. They distract the viewer's attention, and they sidetrack or short circuit the viewer. Sometimes the movements call so much attention to themselves that the viewer has lost the main intention of the message; thus, it is generally assumed that a performer remains perfectly still until the performer is motivated to move.

Position

Central to television and film work is the moving performer and the corollary, the moving camera. The camera moves more often and further, as a rule, than the performer; therefore, a performer much of the time is motionless. Nevertheless, a performer makes contact with viewers by looking directly into the camera lens, such as in newscasting or introducing a guest. At other times the performer avoids the lens and looks at other persons on the set, in which case viewers become observers. Most of the time viewers are observers.

To enhance audience contact, you must keep aware of the camera and play to it, called *cheating*, by maintaining open or closed positions relative to the camera lens. An *open* position allows the lens to reveal your face and body easily. The director can get the best shots of the talent, as you tend to face or *cheat toward* the camera. For instance, as you walk, open a door, or hand an object to a guest, you do it with the limb farthest from the camera, so that your body position is open. During an interview, instead of facing your guest directly as you might in normal conversation, both slightly favor the camera lens in their gaze and position. A *closed* position prohibits observation. Some performers may compare the performer-camera lens relationship to the basic performer-audience positions on stage, except that they are in reference to the camera lens: directly facing the lens (or audience), *one-quarter* turn right or left away from the lens, profile, *three-quarter* turn from the camera, and full back to the lens. *On-camera* means you are actively in the scene and are probably in camera view. At such times you always conduct yourself as if in camera view. Strictly speaking, you are *off-camera* whenever you are out of camera view, even though you may be a participant in the scene.

Naturalness

Children move naturally. As a person grows older, he or she may become aware of social restraints. Social conduct imposes certain movements and gestures upon us. You become conscious of the way you move. Self-consciousness retards the natural ease to which you once were accustomed. Because of the self-conscious movements developed in adolescence and beyond, most performers must be taught how to move so that they appear natural, yet graceful, before an audience. Common movement may be quite difficult: walking across a room, sitting in a chair, moving your hand. Society has dictated what is considered graceful, and you must learn what is accepted. Often your first awareness of your movement occurred when you appeared before a group. Suddenly you are pulled from the anonymity of the peer group, and are singled out by being placed in the spotlight. Fear strikes and you may be temporarily paralyzed. Hence, the common expression, "I froze." In any case you cannot move with abandon until you learn how to be natural in front of a group or on mass media.

Naturalness is the appearance of being at ease, warm, friendly, and gregarious on-camera and on-microphone. Naturalness in a performer varies with age, experience, and self-confidence. Custom and education tend to dictate behavior. Education moves an individual in the direction of the norm, the socially accepted standard. Peers tend to force a person toward demonstrations of common practice in regard to appearance and action, resulting in the reduction, redirection, and elimination of individual qualities. Nowhere is this tendency more evident than in news. News reporters appear similar. Usually they try to emphasize content and not personality. Consequently, most newscasters downplay appearance, movement, and style.

Posture

Ideally, a person grows up tall and straight, as if a thread, attached to the center of the top of the head, continuously pulled that person skyward. The tilt of the head should allow the eyes to look directly forward, the shoulders should be level, the chest should be expanded to the extent that it looks developed and capable of taking in a good supply of air, the stomach should be flat, the upper torso should rest directly on the hips, the buttocks should be tucked in, the body should be supported by straight legs (without locking the knees), and the feet should point directly forward. The entire figure depends on a backbone that is essentially straight, allowing for a slight curve inward at the lower back. Your posture can be quickly checked by standing next to a perpendicular surface such as a wall. What parts of your body protrude? Typically, a novice may lead with or thrust the head forward when walking. The chest may be sunken, the stomach and buttocks may stick out, the legs may be bowed, and the feet may point outward at 45-degree angles. As you look at your posture in a mirror or on a videotape, observe these potential problems. If these problems exist, retraining at the conscious and subconscious levels may be necessary. Of course, physiological peculiarities may limit an individual from precisely achieving these goals.

THE BODY IN MOTION

Standing, sitting, walking, climbing stairs, falling, and bowing are the principal movements required for television performers. The objective is to move under control and appear relaxed, despite being the center of attention. Your stance should be energetic, your walk should be fluid, your sitting should be comfortable with the body positioned slightly forward, your ascending and descending stairs should be effortless and without peeking at the stairs, your falling should look realistic without hurting yourself, and your bow should be standardized. (Pick out one you like and keep it.) Everything you do should be rehearsed until it is exactly right. For now go through these movements before a mirror or have them recorded on videotape, or both. Here are the items to note, for these are frequent errors: Do you fidget or move slightly out of position while standing? This is known as *drifting*. Do you swing your arms, stoop, or move your feet like dead weights when walking? Do you sit down with a thud, and then collapse in a sprawled position? Do you wobble at the ankles (especially in high heels) or appear otherwise shaky when climbing stairs? Other deficiencies are insufficient practice in falling (as in gym class) and uncertainty in bowing.

A great deal of detail can be discussed for each of these basic movements. Walking, for instance, has direction, strength, duration, speed, timing, and grace. Direction is relative to the camera you are on, but in television many cameras are used; so you can be moving right to left in Camera One, and left to right on Camera Two. Whatever direction you are going in, you will do so with a degree of strength related to your motivation, perhaps ranging from lackluster shuffling to running. How long the movement lasts on camera, whether it is in slow or fast motion, whether it is awkward and unpolished, or fluid as a superb athlete, suggests the possibilities that you have in developing your abilities. Remember, if you are role playing, movement that would be ordinarily considered in error could be correct. Once you take your work seriously, you will strive toward perfection just like a champion or leader in any field.

Balance

Animating a static pose requires balance if you are to have facile, fluid movement, instant bodily response, and quick changes of direction. Establishing your balance will give you emotional as well as physical stability.

1. Assume an erect stance. March in place gradually raising knees higher and higher, or brush the right foot forward and backward several times, then the left; or kneel on the right foot and then the left, moving the entire body up and down several times.
2. Mark a straight line on the floor, then walk it like a tightrope. At first you may want to use a rod or chair to improve your balance.

3. Squat to the floor, extending your arms forward. Slowly rise, adjusting your feet to a secure stance. Note the position of your feet and body in a mirror. If the stance is attractive as well as balanced and comfortable, remember it, for it can be used as your basic standing position. Practice moving from this position by taking a few steps in various directions.
4. Walk directly to a chair. Sit in it. Practice moving around the set until you are absolutely secure in your movements, efficient in your direction, and smooth in every gesture.

Animation

In one sense animation simply refers to movement of the face and body; in another, animation gives life or a quality to the movement by making it more interesting, exciting, and impressive. Such movement has been formalized in an art form practiced for centuries called *pantomime*. Pantomime is communicating ideas, emotions, and narratives by means of bodily expression without words. A few basic principles are useful to those in news and information:

1. The chest is the key to all bodily action.
2. Positive emotions are expressed through an expanded body. Hold chest and head high. Keep movement free but conservative.
3. Expression in the face precedes action; thought is first detected in the eyes, then flows over the body.
4. All action must be definite in thought, execution, and motivation.
5. Negative emotions are expressed by a contracted chest, restricted gestures, tenseness, and drawn features. You should avoid this kind of bodily action.

MOTIVATION IN MOVEMENT

Inasmuch as unnecessary movement wastes energy and may blur the message, an information person moves mainly for six reasons: (1) to define purpose, (2) to catch attention, (3) to tell a story, (4) to convey emotion, (5) to delineate character, and (6) to explain relationships.

Purpose

Movement should have purpose. Obvious movement—moving your head up and down to read a script, turning to see a studio monitor, pointing to a chart, holding a dish for a demonstration—are necessary to construct the visual message. A hostess moves from the shadows into a pool of light in the center of the studio floor. The purpose of her movement is to be seen. She knows why she is moving. Frequently, as in this case, the movement is for technical reasons. More often, action is generated from the material.

Attention

Movement catches attention. The smallest movement will draw attention if a viewer sees it. In general, movements are obvious and intentional. You want the viewer to follow your action so that the message is understood or your ability is appreciated. Movement can be analyzed as (1) capturing attention, (2) focusing attention or (3) redirecting attention. To illustrate: a spokesperson stands near a refrigerator on camera right. She opens the door with her right hand (capturing attention). She moves her right hand past food shelves calling attention to space and quality (focusing attention). She closes the door and walks to the freezer side (redirecting attention). How distracting it would be if during the presentation she wiped her nose, looked off set away from the refrigerator, or limped. These movements may seem absurd, but many performers have equally distracting movements which immediately draw viewer attention away from the product.

Storytelling

A salesman walks up to a potential customer considering a car. A second salesman drives up in a competing automobile. The pair grab the customer and engage in a physical (as well as verbal) tugging match. This is physical narration, wherein aural and visual elements work together harmoniously to communicate a message. In another instance, two boys are playing in a park. A woman's voice calls out, "Johnny, time to go home." The second boy is left alone. Preliminary action frames the message read by the voice-over announcer: "Mike doesn't have a home. Won't you give him one?"

Emotion

Professionals usually control their emotions. When a football player throws a ball onto the ground with force and glee, when a political candidate raises her arms in victory, when a news anchor narrates the death of a head of state or colleague killed in the line of duty, emotion is expressed. These examples are probably spontaneous and uncontrolled. Often, emotion is used as a persuasive tool. When movement that conveys emotion is morally or ethically justified remains debatable; it is common in sermons or the advocacy of a point of view. On one public affairs telecast, a woman opposing the Equal Rights Amendment for women turned her chair so that she would not be able to look at her opponent. Most emotion is controlled and subtle; it is secondary to the message. Some fundraising drives showing disabled children struggling to walk or hugging promoters are primary uses of emotion in movement.

Character Delineation

Movement should delineate character. A common exercise in a performance class is to ask each student to walk across a room, a distance of 15 feet or so. Then the instructor asks the class to analyze what it saw: What is the person like? Is the person ambitious, fun, dynamic, nervous, or friendly? What is the position of the head, the eye level? Did the arms swing much? Was the walk straight? From these questions a class learns to observe carefully and to notice that every person is a different character. Careful study will enable you to improve your own image, and for information specialists such study will let them recognize different kinds of movement that apply to themselves.

Relationships

Movement should explain performer relationships. In a newsroom the placement of the reporters around, next to, or near the anchor explains their relationship. Typically, the anchor is in the center position and turns right or left to acknowledge the sportscaster, weathercaster, and contributing reporters. The main person—anchor, candidate, community spokesperson, or authority—always holds the optimum starring position in the center. Such decorum is adhered to on talk shows where incumbent guests move over one space at a time to yield the single guest chair next to the host. In the telecast of a message at a political convention, the president is the star. The supporting persons cluster around him. In an entertainment program the curtain call illustrates the relationship of performers: The star is in the center, featured players are recognized and nearby, supporting players are in the second row or at the ends of the front line, and incidental performers or groups are in the back row. In church ministries and major talk shows, performer relationships are more complex, involving performers moving toward and away from each other to demonstrate attitudes, yield attention, and reveal information.

CHARACTERISTICS OF MOVEMENT

Every movement can be defined and studied in terms of direction, strength, speed, duration, timing, and grace. No movement should signal anticipated action to the viewers, but certain movements may cue the director. As mentioned earlier, perhaps the most important movement in television is the movement of the lips. If this movement is extended broadly, curved slightly upward (direction), with sureness (strength) and quickness, held for a few moments (duration) after a remark or significant action (timing), and done with ease (grace), it is usually interpreted as a friendly smile. The smile aptly illustrates characteristics of movement which can be readily applied to broader actions.

Direction

Direction on television has to do with the relationship of the performer to the camera. A performer moves up or down, horizontally or vertically, depending on the relationship to the camera. Typically, a performer comes into or out of camera view, and moves toward the camera or away from it. Or, second, talent can remain in a relatively fixed position, and the camera operator can move in relation to the performer. This is equally common. Once the general movement of a scene is determined by the director, called *blocking the action*, the performer knows where to move during the production; but while shooting the scene, the director moves the cameras at will in relation to the performer. The director may tell you, "Don't worry about the cameras, we'll find you," for ad-libbed shots during a discussion program or an interview. At this point it might be said you are at the mercy of the director. A third possibility is to have both the performer and the camera moving. This often happens, as in a moving shot at the courthouse, where client and lawyer are walking side by side. Direction, therefore, refers to the moving performer in relation to a fixed camera and lens; or a moving camera or moving lens or both with a static performer; or a moving performer and a moving camera or moving lens or both. Still more complex is the performer as a multi-image on the screen, showing various views traveling in different directions superimposed or on multi-screens.

Strength

Strength is the degree of power in the movement. Strength of movement is largely psychological: How motivated is the performer? What is the emotional level of action?

Emphatic or uncertain, how dominant is the physical presence of the performer in sheer bulk (weightlifter)? In agility (pole vaulter)? Strength is especially important in some sports events. What is real and what is illusion in wrestling, soccer, or hockey? Showmanship thrives. Performers rehearse the strength of the delivery of the movement without real physical contact: a fist to the jaw, which misses in reality; a karate chop to the back of the neck, which stops prior to neck injury. The thrust of the move in each case is at full strength, but only the illusion of strength exists on impact.

Speed

Speed refers to the amount of time it takes you to move your body or any part of it from one place to another. Fast action can speed up your movement so much that the movement may resemble that of an old film. Normal movement is anything that appears to be associated with the movement you see everyday; and slow speed is familiar to television viewers because of instant replay in sports. You judge your movement in the context of the total performance. Speed can be skillfully used to lend variety to a performance.

Duration

The length of a movement can be so painfully long that the audience tires of watching it, or so acutely short that the audience misses the movement altogether. The right length is the amount of time it takes for viewer understanding, comprehension, and aesthetic appreciation. The last flick of a victim's hand may be the perfect touch to end a story of disaster; a close-up of a face held in supplication may be the proper duration for stressing a point during a television ministry; a child's dash down the aisle of a classroom to announce "no cavities" may give the right amount of duration and focus to a product; dragging a body across a floor in slow motion may provide the psychological impact required for a statement of social reform.

Timing

A performer controls movement so carefully that he or she knows when to be at an exact position on the floor, when to pick up a visual, and when to turn from looking into one camera lens to another. In information programs, speakers know when to deliver a line to get maximum laughter or to evoke sadness. These judgments are collectively referred to as your timing. Timing then is your ability to predetermine the moment you will move.

Grace

Grace is the ease of movement and its appropriateness to the total action. Indeed, some movements are awkward; and they are supposed to be so. Moving with confidence, as though you had done the task many times before, and yet are fresh about doing it again, is thought of as graceful in an ordinary way. Human beings have a natural grace in walking, standing, and sitting that is quite acceptable if it can be preserved for media. Sometimes you experience intense self-consciousness on camera which stiffens the muscles and produces a high level of nervousness which prohibits your work from graceful movement, and you look uncomfortable. Nowhere is this awkwardness more apparent than in high school speech contests in which a student takes two steps laterally or vertically—one, two—to give the student, mechanically, a transition by means of movement. Your uneasiness may be eliminated by a clever director who reassures you that you are doing well. This may be the most important contribution a director makes to a performer.

Involuntary Movement

Common involuntary movement is at the subconscious level. Some part of your body moves and you are unable to stop it. Facial distortion may be the result of frowning, grimacing, sniffing, mouthing words, pouting lips, enlarging nostrils, squinting, blinking, turning down the corners of the mouth, biting the tongue or lips, chewing without gum, picking or scratching the face, leaving the mouth open and licking the lips. Unattractive positions and movements of the hands are clawed, stiff fingers; heaviness of the hands, playing with jewelry, paper, or pencils; tapping or shaking objects; repeated gestures; rhythmic gestures; and wringing the hands. Distortions of the torso are wriggling shoulders and hips, muscle spasms, heaving chest or abdomen, humped back, stiffness in posture, and twisting or swinging in a chair. Distortions of the legs are swaying to and fro, foot jiggling, and locking your legs at the knees. Frequently these actions are caused by stress. Facial reaction registering disgust or disappointment may be particularly unattractive and should be kept off camera. Controlling involuntary movement is accomplished by (1) recognizing the offending movement by seeing it on videotape or being told about it, (2) knowing what action you will take to avoid it, (3) remembering the offending movement when you are under stress on the air, and (4) keeping the offending movement at your conscious level until it has been completely eliminated.

7

Discipline

Mental emotional, psychological, and physical good health are essential to the life of a performer. The objective is to gain through experience and exercise complete control over your mind and body. To accomplish this you must have discipline, which includes a positive attitude concerning your self-worth, ambition to be a competitor, mental concentration, control over nervousness, a reputation for dependability, physical discipline, and longevity.

ATTITUDE

Describe it as you prefer—drive, yearning, obsession, or a *calling*—this is the nature of the attitude you must have to succeed. Few performers succeed by accident. Biographies of performers show a deep, intrinsic desire to appear before the public and to mobilize all assets toward that goal. Sometimes this desire is labeled as having a positive attitude or simply believing in yourself. Frequently believing in yourself is not easy. Performing professionally is after all a long shot, and discouraging data and testimonials abound. Yet, many potential performers have not been encouraged because those persons closest to them have no experience with the field; and, indeed, they are afraid that the field will be bitterly disappointing, a conclusion often based on hearsay and half-truths.

A dedicated performer depends on intuition. With drive, good training, and opportunity, a real performer knows deep inside that he or she can be a success. The signs are evident. For instance, a real performer doesn't have to be coaxed to practice. There is neither applause nor money for such solitary work. If you are a dedicated performer, you will practice until you know the work is good enough to show the public. Journalists do not waste time wishing they could be on the air; they practice writing and speaking until they are fluent, no matter how long it takes. The true performer makes no excuses for a weak show. It is analyzed in detail, and then flaws are eliminated through hard work. If you are a journalist who has difficulty pronouncing certain words aloud, you get a fine pronunciation guide and practice until the faulty words blend into the rest of your vocabulary, and flow with your delivery as smoothly as the most familiar terms.

Belief in yourself, desire for perfection, and self-assertion in practicing your talent form an attitude that builds success. Lacking the right attitude automatically portends defeat, even though you may be very attractive and have a wonderful voice, and your friends say you have natural aptitude for the work. Performing is very difficult. To reach your goal, you must have a dominant attitude that refuses defeat on the way to success.

AMBITION

Ambition is the desire you have to reach a goal. The goal, of course, is to be a television performer. But your ambition must be further refined in terms of the degree of fame and fortune you want to strive for. Do you want international fame or is local recognition more desirable? Is great wealth important or is an economically comfortable life sufficient? Your ambition should be defined, because to obtain your goal, whatever it is, you must pay a price, usually in terms of hard work, sacrifice, and doing many things you don't like to do. For example, national information specialists fly a lot—often over 100,000 miles a year. Many are away from home and family for long periods. A major network news anchor once said he was home only 33 days in his first year with the network. Their lives are under constant scrutiny by competitors and others who thrive on taking advantage of any professional or personal mistakes they make.

By contrast, working at a local station as a community affairs director, where you may have a program of your own, may give you local recognition, reasonable salary, opportunity to be at home, and rewarding feedback from members of the community. You must weigh your values and priorities when you seek work as a performer.

Ambition varies in individuals. Roughly categorized, the most ambitious are obsessed. Whether national figures or working locally, some information specialists are incredibly driven toward professional accomplishments. You must be aware of such competition. A second group consists of those who have an ability to succeed at whatever they do. If circumstances permit them to become performers, they are successful at that, but they are confident they would do well in other occupations, and that they would enjoy them also. A third large group of performers is made up of specialists in subject matter. They may not be distinguished as performers. In fact, comments may abound, "How did he get on TV?" The answer is, he is respected for knowing his business. Expertise in a specialty may lead to a feature series, such as short items on the history of your state, or occasional appearances, such as the station's paid pollster consultant during political campaigns. Media performing whether in broadcasting or on cable television has a wide range of opportunities; therefore, performers should consider the nature of a media career in the context of enhancing their lives. Ambition should be measured against the values the performer hopes to receive including self-worth, community contribution, salary, medical benefits, profit-sharing, stock options, retirement, and other items that all performers eventually take into account.

CONCENTRATION

Concentration simply means paying attention to what you are doing to the exclusion of other elements. Television studios abound with distractions, but you must focus attention on the information you have without interference from nonrelated activities and stress. Concentration enables you to become absorbed in your contribution to the program. Some performers are much better at concentrating than others; some people say they can concentrate while listening to a radio or television set. Others require absolute silence. Whispering or laughing by crew members can be especially disconcerting if you think you are the target of unattractive remarks. If the studio is noisy, bring that to the attention of the director, discreetly, if necessary.

The loss of concentration frequently causes forgetfulness, stammering, the delivery of the wrong line, or losing your place on a cue card. Television and film productions often begin with the famous line, "Quiet on the set." The response to this line focuses attention, avoids unwanted noise, and assists performers in concentration. After quiet is called for, no one fools around, and keeps a job for long. Concentration can be broken, too, if items, such as props, are misplaced or furniture is rearranged, thereby disrupting the performer's anticipated routine. Expert performers ad-lib around unexpected changes without notice to viewers, but a wise performer checks all props and equipment before shooting a scene. A television program may be shot before a small crew or a capacity-filled stadium, so the potential for losing your concentration is great. You must play to the audience whether it is one person at home, or thousands, by maintaining confidence in yourself, in your material, and in knowing what is being said, and what action is taking place each moment.

NERVOUSNESS

A major deterrent to success as a performer comes from nervousness. Nervousness can be so severe that it prohibits a person from doing acceptable work; it may even jeopardize your health. Nervousness is caused for a number of reasons—fear of failure is probably behind them all. This fear is based on insecurity in your look, voice, material, and memorization. It frequently occurs when a performer is originating a new work. If you could be fully confident that everything would go well, perhaps you would not be nervous before the cameras; but part of being a performer is the risk, the unexpected. You could be a bomb, or you could be dazzling. So you want some nervousness that is generated from genuine excitement. Meanwhile, you work for total confidence in what you are doing, and concentrate on doing it well. Some performers have conquered their nerves to this extent. Usually they are experienced; they have been through every disaster and survived. Experienced performers often reveal how nervous they are. Some performers prefer television and film because they know the scenes can be reshot until they are perfect.

Nervousness tightens your entire body—your face, throat, physique. So you fight to control it. How difficult is it for you to relax? The crew will encourage you to relax, but you are the one who is in the spotlight, not the crew. Nevertheless, if you are to be a good performer, you must assess your tendency toward nervousness, and develop techniques that stimulate relaxation. Relaxation is as necessary for body movement as it is for voice control. A performer must learn to relax. Relaxation is not easy, and tension can be so intense that a performer feels like exploding. Instead, this energy needs to be channeled into the performance, and released in just the right amounts to insure swift pace for the program. Relaxation enables a performer to knowingly control the voice and body.

Determine whether you get nervous before, during, or after a performance. Are you afraid of making mistakes before the cameras? What is the cause of your emotional instability?

DEPENDABILITY

Dependability sometimes is more highly prized in a performer than unusual or exceptional talent. Naturally, a manager does not want to have to choose. Both talent and dependability are desirable. A dependable person who is competent, but perhaps not so highly gifted, may endure longer than a more talented colleague. Dependability means having the work in on time and working in anticipation of extenuating circumstances. The key to dependability is commitment. A performer, regardless of illness, bad weather, family disaster, personal pleasure, or unexpected circumstances, is often on the job because of his or her dependability. Such a commitment is part of the discipline of a performer. A dependable performer is prepared for all eventualities, and this is possible only if the performer can anticipate any circumstances that may arise. Dependability further means that while a producer depends on you, you depend on yourself to get the job done, even though you may accept gratefully the assistance of staff and crew. When you step into the spotlight, you are the one who usually takes the credit or the blame for the quality of the program.

PHYSICAL DISCIPLINE

Every performer should be in an exercise program that is enjoyable and effective. Weightlifting and related forms of conditioning and dancing are especially useful. They enhance your muscle tone, shape, grace, stamina, balance, breathing, rhythm, and overall fitness. Consistent exercise instills consciousness of your body that tends to prevent over- or under-weight. If good health is given a high priority in your life, physical discipline will discourage or eliminate smoking, drugs, alcohol, and excesses that are debilitating. Actually, a performer is like an athlete during the best years of his or her career; but, unlike an athlete, a performer's career may last a lifetime. Regular exercise reinforces consistency in the professional workplace, and exercise does wonders in reducing nervousness. Many performers exercise before they go in front of the cameras to reduce tension and to limber up.

The best place to exercise is in a large room with mirrored walls, so you can see your body from every viewpoint. Allot yourself a certain amount of time for exercising. One hour 3 or 4 days a week is minimal. This brief period is sufficient for tonal maintenance. If you want to accomplish more, such as weight loss or muscle building, you must stay longer and work harder. With many trim athletes of reputation seeking roles in media, aspiring sportscasters who have not made a reputation on the playing field must be outstanding in appearance, also.

Begin your basic physical conditioning with a few stretching exercises. Count each movement, so you can noticeably improve. Record your numbers on a chart; the data will help in the long run, because you can determine your improvement. After limbering up, move to those exercises that are likely to improve the parts of the body you want to work on. Stay with exercises that do the job, that you like, that show improvement, and are rewarding. Eventually you will have a routine that consists of stretching on the floor and at the bar, static bicycle riding or running, weightlifting, and other machine activity that will shape specific areas of the body. Dancing, aerobics, swimming, fencing, boxing, karate, and other sports may supplement your basic program. Try to develop skills at sports that may be relevant to performing on television or film. While physical conditioning can be fun, the routine must be adhered to religiously, and may become dull. To avoid boredom, enlist the assistance of a partner. You can encourage each other to be enthusiastic. Exercise is hard work and requires good discipline. On a note of caution, be sure you use weights that are appropriate for your level of achievement. Excessive weight can cause serious damage to your muscles and joints. Increase weights only when your body is ready.

LONGEVITY

Someone may say, "I can do as well as he does." Yet can he? Frequently this comment disregards the fact that the subject stuck it out through good times and bad, low salary and high, criticism and praise, and adroitly kept on the good side of management. That performer is a disciplined professional in that he has stuck to the business, and, as in most businesses, opportunity usually arrives if one is not too impatient.

Audiences are capricious; they seldom know what they want. Often they simply want change, so performers find work in media unstable. One day a performer is a hit, the honoree of several awards, but before the ceremonies are over, his or her program series is cancelled. Once again you must draw upon your abiding discipline as a performer, believing that tomorrow or soon another opportunity will come along. Trends and styles in media are usually short-lived and constantly changing. Program ideas spiral forth, tending to repeat material, formats, and public concerns every few years. You must be ahead of the trends, always anticipating changes, and striving for improvement and freshness. Staff turnovers, new program formats, new sponsors, and new voices and faces are a few of the reasons an opportunity to perform will open up; and someone with longevity, well-liked, and known to management, often gets the nod.

8

TALENT

Talent is defined as whatever ability a performer possesses naturally. By definition nearly everyone has talent: a gift of gab, a winning smile, sex appeal, good rhythm. For you to be considered talented, however, you must recognize and use these abilities—fluent speech, putting people at ease, funny expressions—for the purpose of exhibition. Talent is a natural combination of abilities that the public wants to see and hear, to the extent it is willing to pay for the opportunity.

INTUITIVE ABILITY

Intuitive ability is manifest in several common descriptions: writing, vocal, artistic, intellectual, and mechanical ability. Some people are fortunate in having abilities that they have developed early to a high degree—a child prodigy who has read classic stories since age five, a young Thomas Edison who has been a science fair winner, a superbly skillful youngster who can design and sell clever items she makes out of scraps around the house; a high school athlete who has been offered scholarships to the top universities, a youngster who has inspired churchgoers with sermons, and an intellectual genius who is invited to join advanced study programs. These people are talented, and their abilities were recognized at a young age. Often these youngsters are misfits—introverts who have difficulty working or playing with their peer groups in school; yet, if their talents are properly cultivated, they may be performers sought after by the media. Media caters to the unique individual.

Improving your ability is the task you and your teachers face, and may be accomplished as much by fate as by design. The abilities of most people go unnoticed, because most people are preoccupied with learning career skills that will enable them to participate in the obvious work force. The lack of time and money are big inhibitors of talented people who find that they are well beyond their youthful years before they can pursue the work of a performer. The perfecting of your ability and the establishment of contacts to exhibit it may take years to develop. Nevertheless, this process can be accelerated through obtaining an education at good professional schools, colleges, and universities.

Latent ability may never surface because the social pressures surrounding the normal desire to seek a place in society will not allow sufficient opportunity for nurturing such ability. Occasionally a superior coach, talent scout, producer, or teacher may spot latent ability, develop it for a considerable period of time, and have it flower. For the most part, you must develop your own talent.

IMAGINATION

An attractive blond did precisely what she was told on camera. She had her visuals in order; her script was ready. She produced her part of the program thoroughly. To all appearances she should do very well in mass media. She seemed to have everything. When it came to planning future programs, however, and the details of her role in them, the director had to come up with clever ideas for her to use. In other words, the attractive blond had inadequate talent for the job. She could not fuse her past experience with current ideas to form a new, original, exciting, fresh program. She lacked imagination. She was always asking, "What do you think I should do next?" Then she would ask, "How do you think I should do it?"

An active, vibrant imagination is part of your intuitive ability. Imagination springs from the innermost self, a creative region wherein the artist, entertainer, thinker, scientist or journalist seeks the truth, and makes his or her appeal to the public. This region within yourself is like a bottomless well from which more ideas flow than you have the opportunity, time, or money to carry out. Imagination can be encouraged, stimulated, challenged, prodded, or suppressed, depending on the working conditions or learning environment in which it is exposed. A good school will let you articulate whatever ideas you have, however daring; and, whenever possible, will help you see those ideas materialize in productions. But first, you must dream the impossible dream, and then decide how to produce it. All top-notch performers have the imagination to dream dreams; a few find the wherewithal to produce them. To improve your imagination, you should do the following:

1. Dare to think anything you wish and express it.
2. Remember that those thoughts may be very valuable someday, although you may not be able to use them now.
3. Know what is happening in media by viewing, listening, and reading a great deal.
4. Gather ideas from art galleries, exhibitions, lectures, sports events, and entertainment of all kinds.
5. Try new experiences through people and activities.
6. Play games of improvisation and story telling that relate to media.
7. Keep a notebook and video camera that you continually fill with ideas and images, no matter how crude or incomplete they may seem at the time.

CHARISMA

Charisma is the ability of a performer to intellectually or emotionally (or both) reach and touch or stimulate and enchant another person, and through media, perhaps millions of people. Some performers have greater charisma in intimate settings; some before throngs. A performer's charisma is conveyed by his or her presentation. A newscaster, for instance, who addresses a camera lens is speaking before multitudes, and is accompanied by films, videotapes, slides, music, and an expensive array of assistance.

People the public likes or hates, but wants to know about, have charisma. Charisma is sometimes called *magnetism*, for it enables a performer to attract the public like a magnet attracts iron filings. Charisma is hypnotic, powerful fascination that an audience has for a performer. Charisma is innate; a performer is born with it. True, the characteristics of charisma can be enhanced through training, but basically it is naturally within a person in varying degrees. In most people, charisma rests at a rather low level, but it is certainly a characteristic of one's friends and neighbors. In everyday discourse, for instance, most people nominally arouse those with whom they come in contact, but they do not stimulate them to an extraordinary level. A performer distills this power and exhibits it under optimum circumstances, and does so in an acceptable, pleasurable way so that the public is continually enchanted. Even if the public becomes disenchanted, however, the performer with charisma is likely to return in time to public favor as the irresistible prodigal or black sheep. The public cannot resist charisma.

Typically, a performer's charisma plays upon a specific response in another individual. Thus, the public may be fascinated by a streetwalker's haunting, pathetic, depraved appearance in a documentary or may be spellbound by the scientific magic displayed by a teacher in a television class. Whether emotionally or intellectually appealing, a performer instinctively relates to the audience in an enticing manner, assertively and purposefully arousing the audience. Charisma is charm: a spell that is believable, at least during the course of the performance and often beyond, accounting for the tendency of the public to associate a celebrity's role as a performer with his or her personal life. Television enhances charisma. Qualities that bring you to public attention are refined and amplified. These qualities, as amplified in media, appeal to the secret desires of great segments of the public and constitute charisma. Every performer has certain qualities an audience likes to watch: romantic adolescence, the heroics of an average citizen, staunch independence, or the aloof charm of the beautiful. The desire of the performer to exude these qualities and for the public to receive them is a shared experience between the performer and the public. That communication, or shared experience, constitutes charisma.

CREDIBILITY

A young child watching television asks his mother, "Did this story really happen?" "No," she says, "it is make believe, and those people are actors." Within a short time, the child will be able to judge the credibility of a television performer for himself. He will eventually adopt a wary attitude, for example, if he finds that toys shown on television do not live up to the announcer's claim. He will learn which adults to believe, and why—this is credibility. To purveyors of information, credibility is essential. Doctors, lawyers, politicians, military officers, and others must maintain public trust if they are to be effective. The bond of trust which the public has extended to professions and to businesses somewhat indiscriminately has been dissolved in favor of a specific trust in some performers. That is why these performers have become so extremely valuable and highly paid. The public believes them.

The basis of a credible image on television is in the look, the voice, the manner, and the message that conveys honesty and rationality. Credibility presupposes at least a modicum of intelligent judgment. Sometimes, a performer can be rather unsophisticated in language, appearance, or manner, and yet be believed by television viewers. To the public, trust in an individual and trust in his or her media image may be different, but not inseparable. The confidence that a politician may generate while sitting in a voter's home drinking tea and discussing the issues of the day, shaking hands or delivering an after-dinner speech, creates an illusion that the voter is in some way close to the politician. This impression must then be conveyed through television so that it can be disseminated to vast audiences. The politicians' speeches are written in simple, easy-to-understand terms. Public officials, furthermore, dress conservatively, unless they wish to use an item of clothing for image identification (such as a bow tie). Politicians move gracefully, smoothly, and swiftly, as though they are in a hurry to meet as many people as they can. Though warm and energetic, they may or may not go in for physical upkeep (facelifts, diets, and so on), but they keep physically fit and sometimes boast about it—for example, rarely wearing an overcoat even in cold weather. Politicians learn to smile on cue, and to be aggressive in shaking hands, such scenes being very popular on television.

SEX APPEAL

Sex appeal holds an inordinately prominent place in society; it is, therefore, important to anyone who utilizes television. Characteristics of sex appeal for men and women vary somewhat, with men seeming to rank visual over auditory enticement, and women the reverse. Both notions are, realistically speaking, largely fantasy; but then, television is not reality, even in its naturalistic moments. Media are illusion; media are light and shadow; media are impressionistic.

News and information specialists, like entertainers, must consider maximizing their attractive characteristics and, yes, sex appeal. Information programs, including most regular newscasts, are a mixture of journalism and show business. Few anchors at any geographical level are unattractive. They seem more or less powerful, wealthy, and sexually attractive. They wear make-up, have their hair styled, appear in a different outfit daily, are lighted and miked perfectly, and look slick and smiling. Most anchors appear to be youthful, if not young. At a station in the city of New York, a female co-anchor kidded the handsome, blond weathercaster on air by saying, "We forgive you for the wrong forecast yesterday—because you're so cute." High profile information specialists are sexually attractive; and, therefore, they are among the much admired members of society, locally and worldwide.

Although physical attraction heads the list when it comes to sexual appeal, action, disposition, and style reenforce the public's perception of an information specialist. Most performers want to enhance their attractiveness to the public, so they engage in activities that show off their abilities and concerns. In the name of charity they exhibit their personal prowess (race cars, ride in rodeos), leadership (host political meetings), public concerns (moderate programs concerning major issues), and playfulness (anchor holiday parades and celebrations). These efforts subtly extend their public admiration and viability or sexual attractiveness. Because of the flexible nature of many talk show formats, performers can show off their less formal attributes by participating in fashion shows (wherein they wear some of the clothes) and demonstrations (wherein they do the feats along with the guest). Hosting one's own program is the ambition of many, if not most, novice information specialists.

Because sex is an integral part of human relationships, and television singles out certain individuals to express it, news specialists as well as entertainers can make a positive contribution to the viewers by their own example. Rather than expressing sex as enticing and overt, such as the demand might be on an entertainer, the information performer interprets sex as vitality, wholesomeness, and positive self-worth. To illustrate: one feature reporter who is over 65 years of age openly discusses the need for a person to maintain himself. He is physically fit (plays golf every weekend), dresses impeccably (has a clothing business), relies on cosmetic surgery and a toupee, maintains a demanding schedule (frequently appears at hospitals, homes for the elderly, parades) and keeps mentally active by reporting on vital community matters concerning the underprivileged. He has also been married to the same woman for over half his life. The public seems to approve of an older person who takes the trouble to cleverly maintain his or her physical appearance; hence, the flood of celebrity books on fitness and diet. The zeal for preserving a healthy, youthful look prevails in many people, especially performers who often manage to survive for many decades, never retiring even though they tend to slow down. Many elder statesmen have kept themselves and their television images through careful personal upkeep, to the benefit of themselves and the public. Such positive aspects of one's life create the perception of sexual appeal. This kind of sexual appeal is highly appropriate to the news and information specialist.

THE SEXUAL IMAGE

Your sexual image is undoubtedly related to physical attractiveness. A pretty woman and a good looking guy have one ingredient for sex appeal; but this is far from enough, for your personality must be pleasant. You want audiences of great variation in preferences to invite you into their homes by means of television. Sexually attractive images are more or less creations, even for those in news and information. Basically, both sexes appear to be intelligent, mature, good looking, and articulate on a wide range of subjects. Their personalities seem to be very likable. Yet, it should not be amazing to learn that these individuals off-camera, behind the scenes, are human beings, too. They have their bad moments, as so many reporter biographies reveal. But as information specialists, they realize that their images must skew toward the conservative side. Suggestive appearance, movement, or innuendo would be inappropriate and even potentially damaging to their careers and professions. Some news contracts go so far as to prohibit appearances in public that management considers contrary to the television news image. For instance, a news person during off-hours may not become part of controversial plays or demonstrations, may not be a nightclub vocalist, and may not engage in activities that draw unfavorable press attention. Such activities, even though they may not be personally harmful or illegal, may nevertheless seem inconsistent with the leadership responsibility of an information specialist. Certain restrictions imposed upon a performer's personal life may indeed be a price that is paid for high visibility in news and information.

UNUSUAL TALENT

A performer's talent may not be confined to ability, such as the accomplishment of an extraordinary task; or limited to a rare designation, such as a world champion.

More likely, the performer may be unique because he or she is part of, and speaks for, a group. This performer can articulate the special world or group that most people do not understand or even shun. Specifically, racial, religious, or sexual minorities, the medically disabled, veterans' groups, the elderly, and crime victims, among others. As the media serves more diverse interests in the future, spokespersons for these groups will be needed. In the past the media has depended on those already on the information staff to identify and to lead discussions concerning the special views and experiences of such groups. Their demands require more than most media generalists, such as community affairs directors, can address. Thus, unique individuals with proper backgrounds will emerge, and they will be presented to the public in a positive way. Some current media celebrities are already out front. Entertainers have been effective in displaying that being black, Jewish, or mentally disabled has many complex dimensions. Veterans of all wars, especially Viet Nam, have come forth on media to receive overdue honors and to articulate their roles in society. Homosexual politicians are presenting their views. Wheelchair-bound individuals have appeared on media in commercials as capable persons, and, consequently, demand that their value be reassessed in the workplace. Similarly, the elderly have come out of retirement to show that they can help instruct the masses who cannot read. New immigrants from Latin America are asking through their own media spokespersons to differentiate Hispanics: Mexicans, Salvadorans, Nicaraguans, and Cubans. As news and information programs refine their content, many more unique talents will emerge and take their places as public celebrities.

TRADITIONAL VALUES

News and information specialists inherit the responsibility of upholding important traditional values in society, as surely as fictional characters in dramas, and perhaps more so, because they are real. For example, a black TV co-anchor learns that she is going to have a baby. As she continues to work professionally every day while enduring the prerequisites of motherhood, she becomes as a symbol of dual achievement to her admiring viewers, especially those in the black community.

Heroes and heroines, both real and fictional, exist at different levels. All of them preserve to some degree society's noblest traditions and values. The performers who portray these figures in dramas or who are, in fact, these leaders in sports, politics, medicine, and religion, serve as an inspiring embodiment of society's noblest traditions. And, because of television, they do so frequently and with great impact. Among a hero's and heroine's highest attributes are (1) a consistent knowledge of and pursuit of what is right and good, (2) an unlimited degree of self-sacrifice for a worthy cause or another person, and (3) an inextinguishable love and understanding of human beings, collectively and individually.

Over the past few decades, television has depicted these heroes and heroines in the throes of an increasingly complex universe confronted with overwhelming terrors. Such a universe preserves the stratification of heroes and heroines, familiar to us since ancient times. Simple stratification consists of, first, god-like, super-heroes and heroines with nuclear, atomic, genetic, or bionic characteristics; second, less powerful, extraordinary mortals, namely, police, firemen, professional people, and sport stars, among others; and third, ordinary human beings who act bravely under unusual circumstances. All of these performers serve in media, and often in their personal lives, as an inspiration, especially to youth.

In today's universe the forces of evil often appear to be inundating the forces of good on the one hand; and, on the other, to depict the world as it really is, performers as heroes and heroines (more often than villains) are revealed in shades of imperfection. Thus, the hero or heroine is presented with a double threat: the legions of evil, and his or her own deficiencies. Survival in this world, which is somewhat paranoid over crises, becomes even for heroes and heroines more of a challenge and a more admirable attainment. Despite the controversy over violence in media, it must be encouraging on the whole to have these heroes and heroines consistently meet and outwit the foulest of villains, and thereby reaffirm traditional values of society.

In summary, talented people are often thought of as leaders, as heroes, and as those who embody the best qualities that life can offer. Performers, whether confused with the roles they play or not, are likewise thought of by the general public as bigger than life. If you are fortunate enough to be talented, use your talent to benefit your television viewers, and you will reap untold rewards.

9

The Studio

To a performer studios are merely places where a performance can be originated, recorded, or transmitted under controlled conditions indoors or relatively controlled conditions outdoors. A television or film studio is mostly space surrounded by lighting instruments, scenery, properties, and equipment that will transmit and record pictures and sounds. A performer's primary function is to relate to the microphones, cameras, staff, crew, and possibly live studio audience, while simultaneously communicating news, information, and features to a vast unseen public. Paradoxically, a performer standing in a pool of light amidst the paraphernalia of a large studio (and some are mammoth) can find the experience lonely, frightening, and frustrating. You learn that it is difficult getting instructions from the floor crew or director working on the edge of light; the studios are often cold; your instructions are frequently changed at the last minute; the live audience is impatient; the pressure is on you, and you have no place to hide. Regardless, when you step before the microphone and into the light, you fulfill the efforts and indeed the hopes of the producer, director, production staff, financial backers, and the public.

A television studio is a large room with microphones, lights, and cameras. You must be aware of the placement of the microphones, and the carefully lighted areas designated on the studio floor in chalk or masking tape. These designations are your *marks*. You move only through these areas in order for camera operators to properly align shots. Being off your marks, even slightly, can make a substantial difference in the light level or composition of the shot. You may resent being so restricted in movement, but the less there is to light and mike, the better quality the lighting and audio may be. A typical studio does not have enough sound equipment or lighting instruments to cover the entire floor, and electronic subtleties in television are multitudinous. For instance, if the cameras do not match (an engineer's problem), you may look great on one camera but poor on the other; if the light level drops below camera minimum (a lighting problem), you may be in deep shadow or transmit the wrong color; if the microphone isn't in the right place (an audio operator's problem), you may sound as if you were speaking from another part of the studio.

Television programs may be classified as *live*, meaning they are on the air while they are actually being performed; *live-on-tape*, meaning they are recorded without editing; or *recorded*, meaning they are probably edited. Each method has its values. The material in live programs may be perishable and have little replay value. Live-on-tape programs have spontaneity—as do some audience participation shows, public discussions, or sport programs—with little editing required. The content of these programs may be perishable, also, but they are videotaped for replay at a more convenient time, and for possible syndication. An edited program is extremely popular with performers, because all the content is closer to being artistically and technically perfect, and will be replayed perhaps for years. Frequently, the only live portions of a newscast are of those who anchor it. To be competitive a performer's work must be perfect, and electronic engineering has made that possible. The only deterrents are money and time; the performer's company may not be able to purchase enough studio time to reach perfection; thus, a performer settles for something less. A studio is expensive, because it has the most experienced, creative artists available to do every task.

TELEVISION CAMERAS

A television camera looks like a box mounted on a pole or tripod. Cameras function at the performer's eye level much of the time. The internal workings of a camera are extremely complex, requiring constant adjustment and maintenance. Color cameras are often adjusted by focusing on a chart of color bars or on the complexion of a model. If the skin color, usually a Caucasian, looks natural, the rest of the television picture is likely to be acceptable. Engineering adjustments begin an hour or more before air time. Instruments in master control tell an engineer when the cameras are ready for operation. To you, these technical adjustments may be boring, but they are absolutely necessary if you are to look your best to the public. So, any adjustments that require your presence or your accompanying visuals are well worth the time and inconvenience.

Cameras, their mounts, and their lenses move according to simple commands from the director or camera operators. The principal commands for moving the camera and its mount are *dolly*, which means move toward or back from the performer or object; *truck*, which means moving the camera mount right or left; and *arc*, which means moving right or left in a curve. *Boom* or *crane* up or down elevates the camera overhead. Cameras on tripods are almost impossible to elevate; but those on pedestals move up or down hydraulically on command, *pedestal* up or down. The principal commands moving the camera itself, without moving its mount, are *pan*, which moves the camera horizontally right or left; and *tilt*, which moves it vertically up or down. The variable focus *zoom* lens is on most television cameras. These lenses have a sweeping capability, from a wide shot covering a panoramic landscape to a close-up of an outfielder catching a ball.

At the back of a television camera, sometimes on the studio floor, and in the control room are monitors, or tiny television receivers that allow the director, camera operators, and engineers to adjust focus, lighting, composition, and field of view. Because of the complex nature of television, a director has to be careful that the content of the program doesn't get lost in the complex technical process of getting it on the air. You can help directors, without imposing on their authority, by calling attention to the objects the viewers must see, and when they must see them, in your presentation.

SWITCHERS

Somewhat comparable to the audio control board is the *switcher*. This electronic control panel consists of several buttons that execute changes from shot to shot. These changes can be instantaneous, called a *cut*, of longer duration, called a *fade*, or of overlapping scenes known as a *dissolve*. A variety of electronic transitions, termed *wipes*, enables the operator, the technical director, to use various geometric configurations such as squares, circles, and bars. Additional effects can be inserted into the television picture electronically, such as a tiny window that can move around on the television screen. Colors, distortions, numbers, lines of type, and computerized designs can be added or deleted. You do not have to know how a switcher works, but you should know what effects are available, so they can be utilized to enhance your program.

VIDEOTAPE RECORDING

Videotape looks like audiotape except that videotape is wider—½, ¾, 1, or 2 inches—and the videotape recorders (VTRs) are bigger and more expensive. Broadcasters use all gauges depending on the task. Videotape carries both sound and picture, and can be electronically edited, having flexibility similar to film. You may be familiar with the operation of videotape equipment. Videotape recorders can record, erase, and play back. Operation of the machines is basically simple, but it takes a great deal of practice for one to electronically edit on machines that have that capability. The operation of VTRs is done by a specialist. Videotape is fairly durable and can be reused. Commercial editing of videotape is largely computerized with frame-by-frame, utterance-by-utterance electronic scanning. The potential for perfecting a performer's work through editing is virtually limitless.

STUDIO MONITORS

A television monitor is a voiceless viewing screen. Often it resembles an ordinary television receiver. One or more are usually positioned on the studio floor or at the news desk, within easy view of the performer. They enable you to see film clips for voice-over-picture narration, presentations from other cities, or remote units. Whenever a newscast calls for a pre-recorded segment, the anchor watches it on the monitor. If an emergency occurs or if the picture is substandard, the reporter knows what the viewer saw and ad-libs necessary information. You should make certain that the monitor is within easy viewing range, especially if you are required to identify small objects.

PROJECTION

Moving picture film, slides, and opaques originate from *telecine*, a projection room adjacent to the control room. In projection, various motion picture and slide projectors are arranged so that they beam pictures into a small television camera, which, through a mirror system known as a *multiplexer*, may handle two or three film or slide projectors. You should be acquainted with projected materials: moving picture film, transparent slides, and opaques. Each requires a special projector. Wherever possible, videotape has taken over these functions, as stations attempt to streamline their operations.

At a preplanned moment during a program, a projector may cast a photographic image on a screen behind a newscaster; hence, the term *rear screen projection* (RSP).

Or, the image may be inserted in a box behind the newscaster electronically. Either method may be used. Several slides—primarily of people, significant events, maps, scenes, animation, and even moving pictures—may appear during a single program. Rear screens vary in size from a normal television receiver to a large motion picture screen. Much of what appears to be rear screen projection may actually be produced electronically through the switcher, in which case the performer is said to be *keyed* or *inserted* over the background. This electronic technique is very attractive for announcements, (moving shapes and designs), educational dramas (complex scenery) and news (buildings, arenas). Although the making of RSP slides and electronic effects is left to the production staff, the performer may be involved in the selection of these effects, and should be aware of their availability. Special staging may be required. The performer must avoid wearing blue, for example, because this color is blanked out during the keying, thus allowing the background to replace any blue image.

TELEVISION REMOTE UNITS

In an effort to cope with the immediacy of radio, television news departments have turned to "instant" electronic news gathering (ENG). The process of reporting from various locations outside the studio is referred to as *electronic field production* (EFP). Unscheduled routine events such as accidents and fires are covered by a reporter and photographer who go to the scene in a car. They get the story and return to the station. The bigger events, many of which are scheduled, require a larger vehicle. The EFP truck has a portable camera and audio system. The truck can be set up to quickly relay information to the station, perhaps enabling the newscaster to report from the scene before the event is over or during the on-air news broadcast or both. Preparation is accelerated, however, and the subtleties of controlled studio lighting and other cosmetic niceties are sacrificed. Occasionally transmission is uneven, and you find you must do an unexpected amount of ad-libbing.

Scheduled events of magnitude—entertainment from theaters, concert halls, casinos, sports and fairgrounds, and some public affairs programs— are broadcast live or recorded on videotape by means of mobile television vans. Leased for special events, these trucks contain every piece of equipment essential for putting a complex telecast on the air. The cab area of a modest-sized rig contains a small control room. In the rear is engineering gear, and on the roof a microwave link is set up for transmitting the signals back to the station. Inasmuch as the director is in the truck, the only action the director sees is on monitors. Directors must rely on expert camera operators to pick up shots of the action. The larger vans are accompanied by a utility support van. These units may be capable of using as many as 25 cameras on nationally significant telecasts such as major political conventions, parades, and sports coverage. Some of the most spectacular field work is done from helicopters. A reporter, a photographer, and a pilot may have the only significant view of major traffic problems and disasters.

COMMUNICATION WITH THE PERFORMER

When a performer arrives at a studio, preparation is complete. The script has been memorized or is on cue cards, the blocking has been rehearsed or is at least generally understood. Therefore, you get acquainted with the performing area, and, when appropriate, with special attire or make-up. If no technical problems arise, a rehearsal(s) and airing or taping follow as soon as the director can possibly do them. To meet this schedule, you get a series of check points or *cues*, orally before the airing or taping, and visually during the performance.

Oral and visual cues are given by the director or floor manager. Cuing is precise but informal, unless the circumstances are complex. You must know who is giving the cues, where the person will be, and how visible the cues will be. Once cued, you react instantly and do what is expected by the director. The audience is none the wiser, however. Ordinarily cuing is simple; but in television and film you work through an intermediary—usually the floor director/manager—to talk to the director. Consequently, cuing can be misunderstood or late. In such cases, if the instructions to the talent are simple, the director may resort to the studio intercom (i.e., loudspeaker); but if they are complex, the director will come into the studio to converse directly with you. Do not expect much direction.

A reporter is connected to other members of the television production team through the floor manager and through a headset. The headset, which is a tiny button worn in the ear, enables a reporter to listen directly to instructions from the control room. Many news people feel much more secure wearing a headset. Inasmuch as audibility is all that is required, headsets are not noticeable to the viewers. Most intercommunication systems are controlled by the director, who receives messages from the studio, projection, engineers, remote locations, and elsewhere. The performer may hear all of this conversation.

CUING TELEVISION PERFORMERS

Typical oral cuing prior to the telecast is announced by the floor director. "One minute to air (or tape). . .30 seconds. Quiet in the studio. . .10 seconds. Stand by." A countdown may begin. "Five, four, three, two, one. Cue."

At the 10-second *stand by*, the floor director raises a hand with the flat of the palm toward the talent or points an index finger toward the on-air lens. On cue from the director, spoken over the floor director's headset, the floor director points directly at the performer, who responds instantly. Most visual cues concern timing—the minutes/seconds, wrap up, stretch, or speed up—before the program or segment is over. Some hand signals deal with movement: The floor director waves the performer from one camera to the on-air lens of another, especially if the talent cannot tell from the tally lights which camera is on the air; or the performer, by means of a prearranged signal, such as placing the palms of the hands on a desk in front of him, may indicate that he intends to stand. Some signals have to do with content. If a floor director taps the palm of one hand with the index finger of the other, this signals the talent to go to a spot announcement. If the floor director points to an item on the set, you begin discussing that item. For the most part, so long as you remain on your marks, stick to the script or program outline, and perform as the program was rehearsed, cuing is minimal.

At its conclusion, you should not move until the floor director announces that the show or taping is over: "That's it. Good job. Thanks a lot." Don't move until the cameras have obviously broken to another set. A common error for a novice is to move, especially move your eyes from the on-air camera lens, or grimace prematurely. Such action might ruin the take.

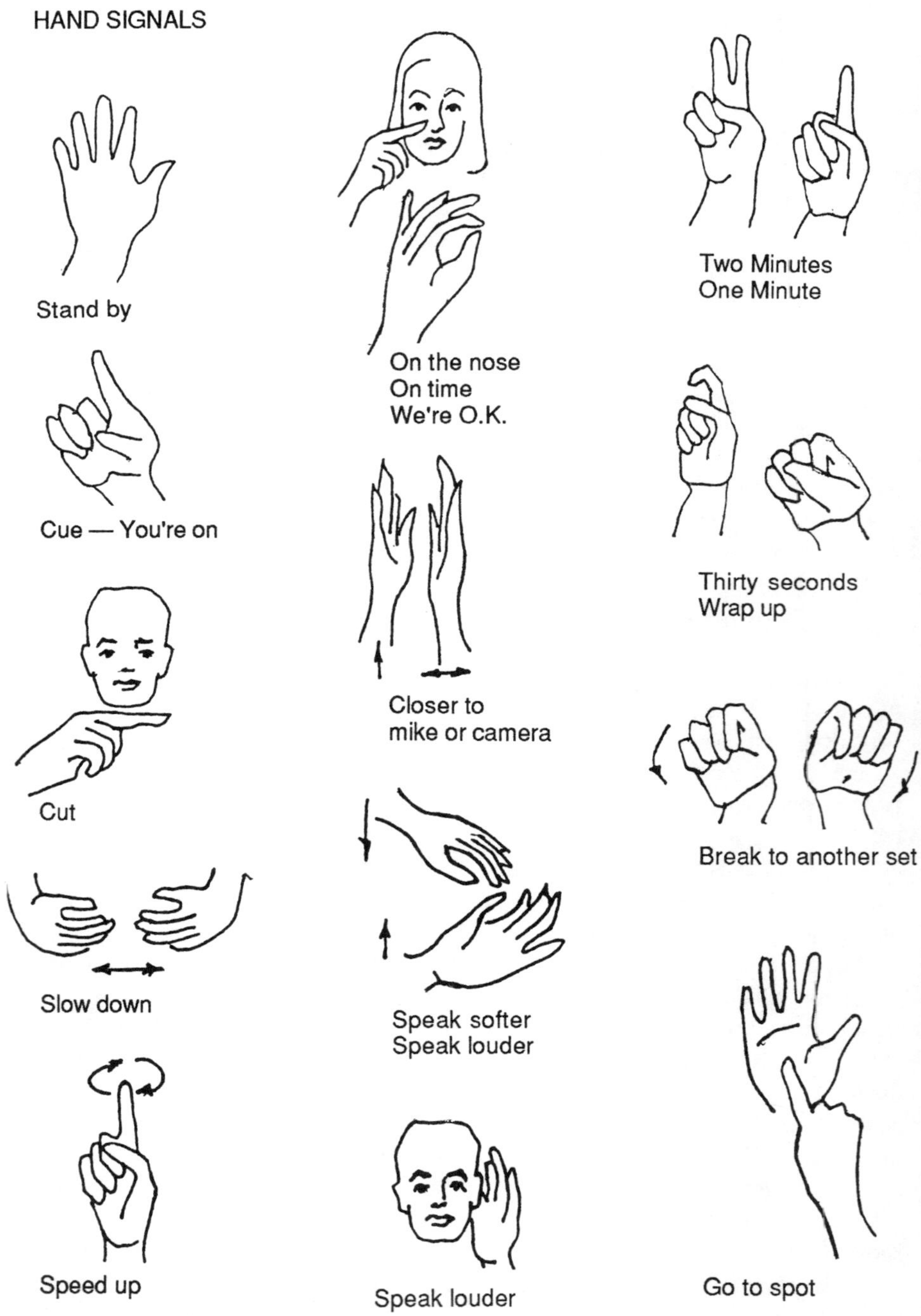

▶ ***Figure 3*** *Hand Signals.*

PERSONNEL RELATIONSHIPS

If a performer becomes a celebrity, it is because of the talent, ambition, contacts, expertise, and cooperation of other performers, and countless people on the production staff. A performer is well advised to be reciprocally cooperative and generous. If the staff likes a performer, it often gives an extra measure of devotion to make that performer look good; but if a staff dislikes a performer, the reverse may result. For instance, as a performer you may not need a rehearsal, but keep in mind your crew may need it. And, if you are unable to stand in for yourself, then hire someone to do it.

AGENTS

Your agent, in some cases a relative, teacher, or close friend—someone who is responsible for you—should represent you in the control room. A person who knows media well can pass along invaluable feedback concerning the performance or the limitations of the production. The agent can help to evaluate the effectiveness of the studio crew. This information may be useful later when you have an option as to where to make commercials, public service announcements, or campaign spots. For all practical purposes, program producers are not likely to comment on your work. If they like you, you'll be hired again. If they don't like you, you may never know the reason.

DIRECTORS

In television, the public is hardly aware of the director's importance, depending largely upon how great the talent is on the tube. In film, you may work with an outstanding director. This experience may be the highlight of your career, perhaps a turning point. Great directors and great performers need each other. From time to time you may question a director's competence. If so, do it discreetly, usually through your agent. Avoid trying to settle differences in front of the crew. Everyone becomes too defensive to be constructive. Remember, the director calls the shots while you are on the air. The director/editor makes all of the audiovisual choices for the program. The director is in charge. Even if you could change something, you would find that changes are prohibitively expensive.

FLOOR CREWS

Remember, everyone on the floor crew counts. If someone fouls up an assignment by being noisy during the take or by making visible mistakes, the program is diminished, and you may be blamed by the viewers for the inadequacies of the production. You cannot afford having thousands of viewers associate you with poor work. Ineptness in the crew may imply that you are inept as a politician, minister, teacher, newscaster, or talk show host. Be nice to the crew and be cooperative.

USING CAMERAS

The camera shows everything in its field of view. Whether you are scratching your leg, adjusting your microphone, or frowning, the camera will show it, perhaps to thousands of people if it is on the air. You must assume you are on the air all of the time during the airing or taping of a program. *Never* do or say anything that could not go over the air.

Cameras are frequently big and clumsy, especially commercial ones. They move slowly from set to set. The zoom lens adds fast and fluid variation, from the wideshot covering the entire set to an extreme close-up of the performer. No matter how many lenses are visible, a camera has only one that is *hot* or on the air. This hot lens is in a different position on different cameras, depending on the model. You should be certain which lens is hot. Often, cameras have tiny lights indicating they are on the air. They are called *tally lights*.

DIRECT AUDIENCE CONTACT

Performers who wish to address an audience directly look into the on-air lens, just as though they are speaking to a close friend. Newscasters, politicians, teachers, and clergy frequently look into the on-air lens. Any other position, such as glancing down to read a script, is disruptive, distracting, and sometimes breaks communication completely. To some extent, an audience accepts the reading of notes because of the tradition of lecturing and public speaking from notes. The illusion of fluid conversation, either ad-libbed or read from prompters without interruption, is better direct communication between the performer and the audience. This means looking directly into the camera and, when necessary, discreetly changing your gaze from one camera to another by looking directly into Camera One, then down at the script, and up into Camera Two. Peripheral vision allows you to see the floor director wave you from one camera to another camera.

INDIRECT AUDIENCE CONTACT

If a performer is addressing someone on the set, the public acts as an observer to the scene. Except for some panel discussions, perhaps, the viewers are less involved. Frequently, interpretive work such as readings, singing or acting allows the viewer to participate in an experience vicariously and indirectly. An actor's emotions and interpretation of a role are observed by the public, but the public is not a direct participant. Instead, you either speak directly to another person on the set, as in a play, or speak to a focal point arbitrarily fixed in the black depths of the studio. Although camera operators move as the director dictates, you pay no attention to their placement. Few narrators talk directly to a camera lens. Most of them talk to a studio audience or to an imaginary focal point. Often the point is about 10 feet off the floor and in direct line with a bright light called a *key* light. This light illuminates the face of the performer, and it tends to add sparkle to the eyes by making them water slightly.

USING MICROPHONES

A performer is concerned with a microphone's frequency, range, coverage, size, placement, and use. The term *presence*, referred to previously, is the degree of fidelity that a sound has. It is the notion that this sound is what an audience would hear if it were in the same room with the performer and listening under optimum conditions. So, engineers do their best to convey optimum presence through media transmission and recording. Your knowledge of recording equipment and its proper utilization will help immeasurably to give an audience optimum presence.

Each microphone has its own characteristics. The principal microphones a performer uses are (1) the dynamic or pressure microphone, which tends to favor the high frequencies, (2) the ribbon, which favors the low frequencies, and (3) the condenser, which gives the truest frequency response. The dynamic microphone is the least expensive and most rugged. It comes in various models. It is omnidirectional, meaning that it picks up sounds from everywhere, whether the performer wants it to or not. The ribbon microphone is much more selective, having a narrow beam extending fore and aft in a bidirectional pattern. The condenser microphone is often used on complex productions, partly because its omnidirectional pickup can be made unidirectional.

The smallest microphones are the size of a tie tack, and may be clipped to a suit or dress. Though tiny, these microphones are usually noticeable, especially on a dress. Somewhat larger is the lavalier, which hangs around a performer's neck on a lanyard or clips to clothing. About 6 inches in length are the handheld microphones, which can be mounted on stands or in brackets suspended from overhead booms. In addition to the uni-, bi-, and omnidirectional pattern for microphones is a heart-shaped pattern known as a *cardioid.* A super version of the cardioid, about 18 inches long and casting a rather narrow pickup pattern some 10 to 15 feet, is a *shotgun*, which is mounted on the balcony of a theater and aimed in the direction of a performer on stage. Wireless microphones attached discreetly, perhaps to the head, have a transmitter attached to a belt. They enable performers to work within a 25-foot radius of a larger receiver-transmitter backstage, thus relieving the performer of an unattractive microphone cable. Their frequency, as well as their range, however, is limited.

MICROPHONE TECHNIQUES

Keeping in mind that a microphone is relatively delicate and expensive, a performer tests it by simply speaking into it, using the same volume and pitch of the voice that you intend to use on the air, and maintaining an appropriate distance from the mike. Tapping, whistling, blowing, and yelling into a microphone for testing purposes are not only unnecessary but may permanently damage the instrument. Ribbon microphones are especially sensitive to mistreatment. The key element, the ribbon, stretches permanently if abused, reducing the frequency range. The engineer will adjust the level of a microphone on the console accordingly, so that every performer's voice blends with the rest of the program. If you are at a loss for words when an engineer asks "Level check, please," simply read the script or count until the engineer has heard enough to adjust the levels on the audio console. Do not stop and start capriciously. Your voice registers on the console's meters only when you are speaking. In general, a performer, especially in a unionized market, does not move microphones. This is the audio engineer's job. Besides, moving a microphone may jeopardize the sound quality or the aesthetics of a shot. If you are not comfortable with a microphone, this should be discussed with the floor director or director. Testing a microphone takes only a few minutes.

A performer works 4 to 8 inches from the screen of a dynamic microphone and speaks across the surface of it, depending upon how much you want low frequencies to register. The closer you get to the screen, the better the low frequencies will register.

But breathing and plosives in the English language will begin to be heard as "pop." Condenser and cardioid microphones also use the pressure principal, but when their pickup patterns are uni- or bidirectional, they are used like a ribbon microphone. That is, a performer speaks directly into them.

Some suggestions for microphone placement may be helpful, as follows:

If you must have a microphone around your neck, place it under your clothing, namely your tie or scarf. If clothing muffles the sound, your engineer will tell you. Fasten your lavalier. If it moves, it may hit buttons or jewelry, even swinging over the surface of a shirt may be noisy. A jacket or patterned blouse helps women, particularly, to draw attention away from the microphone.

If you must hold a microphone while standing, remember the long cable will show. To minimize its visibility when you are standing, attach it to your belt, being certain to leave enough slack so that it will not pull on your clothing. To repeat, nothing looks worse than a performer wrestling with a long microphone cable. Preplan so that you have ample cable to go wherever you must during the performance.

If you are working with a microphone stand, remember to practice in advance if you intend to remove the microphone from its mount. Struggling with a microphone looks unprofessional. If the microphone is suspended on a boom overhead, avoid the performer's tendency to drift out of its beam.

In general, if you are simply speaking, one microphone, preferably unseen, should be enough. If you are a newsmaker, a battery of microphones may be necessary to accommodate each reporter with a sound bite, but this is really the same as one microphone to one performer. If you use a separate sound source, mike it separately, If you are part of a round table discussion, one good omnidirectional microphone is adequate. If you are on a panel—seated in a straight line—probably one microphone for every two people is necessary. Do not attempt to pass the microphone, because it will be noisy. If you are in a group, all principal participants should be miked separately. If you are narrating at a concert, however, the orchestra may be picked up on one microphone suspended in the center of the auditorium, and you will probably have a personal microphone.

TYPICAL SETTINGS

A simple setting consists of the background and floor, scenery or furniture, set decoration to make it look finished, and lighting. Properties (*props*) are items you must manipulate during the presentation. A *cyclorama* is a drape, often made taut, or plaster wall (*hard cyc*), usually white, gray, or blue, that encloses much of the back of the set. Most performers look good in front of a blue cyclorama.

Furniture can be troublesome. Some things you should watch for are furniture that dwarfs you, that relaxes you to the extent that you look slovenly on camera, that is too attention-getting by being too colorful or complex in design, or that obscures you from the audience, such as a large desk tends to do. Molded plastic chairs with straight backs, shallow seats, moderately padded, and of medium gray scale in color are common. They should be armless so that you can get close to your guest, and both feet should touch the floor when you are seated. Most chairs are too low, too deep, too bulky, too cushioned, and too hard; consequently, stools are sometimes used. The stools should be of the right height to give the performer a strong vertical line.

In reference to set decoration, you should be certain the flowers, papers, books, ashtrays, sculptures, painting, and miscellany do not call attention to themselves, nor should they appear at odd places through optical illusions. Plants behind a performer may appear to be growing out of his or her head, or during a newscast, a figure on a rear screen projection may be pointing into the newscaster's ear. A bowl of light flowers in front of dark clothing may be undesirable, just as metal objects may cast unattractive reflections on the performer. For the most part these problems are the responsibility of the production staff, but you should be aware of them. Most sets are disappointing, makeshift, tacky, even dirty and discolored to look at. Do not worry about this. Check what the setting looks like on camera. It will probably look fine when lighted properly. Don't worry about the space you have. Some of the finest programs are done with virtually no space.

10

The Business

For those who want to make a living from performing, some knowledge of the business aspects is essential. This begins with a typical overview of the lifecycle of a performer. As appropriate schooling comes to an end, a performer gets an internship or part time job or both in the industry. Through applications and contacts, a full-time position as a performer materializes. Hardly on the first job, you are already looking for a better one, and 6 months to 2 years later you have found it. You move to a larger market. Your benefits, your abilities, and ratings begin to give you recognized status. Network scouts and talent agents are looking to negotiate future contracts. In a few years you find, if you are fortunate, that you can demand a high salary and other benefits, more or less in direct proportion to the public's enthusiasm for your work. You are now recognized locally, regionally, or perhaps nationally as a media celebrity. In a few cases, some persons have gained recognition by forming their own companies and packaging their own series. A stockbroker hosted her own series on investing. It ran for years on major TV stations. Recently, an appliance repairman made a video cassette series for direct sales to consumers. If the public interest in your performing wanes, as eventually it might, your recognition will enable you to enter other careers like producing, directing, or public relations in media, industry, or government.

Very few people, including performers, know what they want to do at an early age. Opportunity has a great deal to do with this decision. Frequently performers are not ready to take advantage of the opportunities they get. Each professional career in journalism or information requires a different preparation, and so a performer must, as a network anchorwoman once said, get in the right line. If you want to be in news, enter it with a desire to report and to present news. No news director wants to admit hiring just another pretty face, even though a news director may do this to get ratings. If you want to be an on-air presenter, have confidence in yourself and seek your chance to show what you can do. Do not be dissuaded from becoming a performer because someone else says, "There are no jobs." There still may be a job for you. No one can make that decision for you, because no one else can say how much time, energy, study, and work you are willing to devote to performance, or what opportunities will come along. If you know what you want, opportunity will come. Be ready to make use of it when it does.

Typical entry points for television performers are as part-time or full-time, weekend, and odd-hour reporters for television news departments; community or public affairs directors having their own programs; sportscasters, weather people, and farm reporters who can do general news; presenters in closed circuit television who appear

ABC Television Network, 77 West 66th Street, New York, New York 10023

American Women in Radio and Television, Inc., 1101 Connecticut Avenue, N.W., Suite 700, Washington, D.C. 20036

Broadcasting Magazine, 1705 DeSales Street, N.W., Washington, D.C. 20036

CBS Television Network, 524 West 57th Street, New York, New York 10019

CNN, Box 105366, One CNN Center, Atlanta, Georgia 30348-5366

Electronic Media, 740 North Rush Street, Chicago, Illinois 60611

ESPN, ESPN Plaza, Bristol, Connecticut 06010

Home Box Office, 1100 Avenue of the Americas, New York, New York 10036

Medialine, Box 51909, Pacific Grove, California 93950. (800)237-8073 or (408)648-5200. Daily updated voice mail box listings for TV news, production, and promotion.

National Academy of Television Arts and Sciences, 1560 Broadway, New York, New York 10036

National Association of Broadcasters, 1771 N Street, N.W., Washington, D.C. 20036

NBC Television Network, 30 Rockefeller Plaza, New York, New York 10112

Radio Television News Directors Association (RTNDA), 1717 K Street, N.W. #615, Washington, D.C. 20006

Women in Communications, 2101 Wilson Boulevard, Suite 417, Arlington, Virginia 22201

▶ *Figure 4 Job Placement Sources.*

as instructors, announcers, or narrators for education, business, and medicine; some actors and models who are in commercials and host programs; representatives from various community organizations such as the Chamber of Commerce, philanthropic groups, churches, legal and medical professions, stock companies, banks, and trade associations who provide public service series. Cable TV systems and especially public access television systems, along with schools, probably offer inexperienced talent the greatest variety of on-the-air opportunities today, everything from announcing to gymnastic exercise programs.

TYPICAL PROCEDURE

"In order to get a job, I've got to have experience. In order to get experience, I've got to have a job!" This frustrating dilemma is partly true and partly nonsense. With rare exceptions, you must have experience to get a paying job. Getting a job—any job—consists of having an opening and someone else being qualified to fill it. Talent agents, some schools, and a few placement offices tend to know where the jobs are, and they match their clients or students with the openings. The task of matching per-

Entry Level:

TELEVISION REPORTER. Network affiliate seeks general assignment reporter. Degree in Radio/TV/Journalism and experience preferred. Send tape and résumé to: Name, News Director, Station Address. No calls. (Do not call the station) EOE. (Equal Opportunity Employer)

HOST/HOSTESS for daily Entertainment Magazine show in top market. Experience in magazine format and field producing desired. EOE. Respond to Box Number. (at information source or placement service)

RADIO NEWS ANCHOR/REPORTER. Conversational storyteller with strong writing and production skills. Aggressive able to dig and bring stories to life. Experience preferred. Send tapes and résumé to Name, News Director, Station Address. (Do not overlook radio as a beneficial experience)

Next Level (Many more opportunities exist for those with two or three years experience):

REPORTER/PRODUCER: Involves writing, producing, reporting for industry videos. Operate in-house edit system. Television reporting experience required. Résumés only to Name, Insurance Company, Address.

WEATHERCASTER/REPORTER: Experienced, bright energetic person to do weather on weekends and general assignment reporting three days a week. Affiliate with weather computer. Tape, résumé to Name, Address.

NEWS REPORTER: Bachelor's degree in journalism or related field. Experience in writing, producing, editing. ENG experience helpful. Apply to: State Educational TV Authority, Address.

CO-ANCHOR: If you're the best anchor/reporter on the beat and your tape can back it up, let's talk. We need a dynamo right away. Mid-70's Midwestern market. Tapes and résumés only. News Director, Address.

REPORTER: Spanish language network television news. Requires bachelors in journalism plus two years experience in job or as TV news correspondent. Must have network and foreign news experience. Must be proficient in Spanish. Forty hours. Send résumé to Job Placement Company, Address. Job #23456. EOE.

▶ ***Figure 5*** *Typical Job Placement Advertisements.*

formers and jobs is so large and varied that station personnel departments are constantly looking for new talent and replacements; they cannot depend solely on talent agents and schools having the right human resources when they need them. A performer, therefore, should provide identifying materials to talent agents, schools, placement offices, and media personnel.

Letters of Inquiry

Many performers, about the time of graduation from college, blanket the top markets of the nation with letters of application. Some performers have sent as many as 200 letters. This is often a waste of time, because you may receive only one or two offers per hundred letters. More helpful is a letter addressed to a department head at a station in a city which you intend to visit. The letter should be neatly typed and grammatically correct. It should indicate that you would appreciate seeing the station and the department you admire, and that you are seeking the advice of the department head concerning a career in media.

This advice applies to agents and placement services, too. If a response is not forthcoming, send follow-up inquiries and try to make an appointment when you are in town, anyway. Basic information on stations, networks, agencies, and production companies is found in *Broadcasting/Cablecasting Yearbook*; and job openings are listed in the weekly *Broadcasting*. The Radio Television News Directors Association runs a job information service, and so do some instructional and industrial television associations. Placement services may be quite specialized.

Personal Interviews

One news director gets 35 letters of inquiry or application a month. He sees few applicants. To an extent, employment depends on reaching a level of acquaintance by letter or in person. You should try to see the employer in person—the personnel head, the news director—so that you can sell yourself in person. This may take a little doing, because secretarial staff often screens potential applicants. Do not be discouraged. Use relatives, friends, teachers, and mutual acquaintances to get an appointment, if you need them. But get in! Most department heads—news directors, for instance—are very pleasant; and they are a little flattered and greatly interested in new talent who present themselves in a serious way. You should not ask for a job, because chances are the department head will not have one at the time; and thus, the visit would end without having time for you to leave a favorable impression. Instead, ask for whatever help the department head can provide: suggestions, contacts, job leads, or anticipated opportunities. Rarely does a job materialize at the moment of the interview, but sometimes one does later; and the department head, if favorably impressed, will no doubt contact you, assuming that you have the necessary qualifications.

Do not expect the station to pay travel expenses. Be willing to go to the station at your own expense, if there is a chance for employment. Many network jobs have been obtained by aggressive people who were willing to invest in their own future. A department head is quick to notice an aggressive, yet attractive professional approach. Dress appropriately and moderately. Some stations purposefully conduct internship programs giving them an opportunity to look at talent before hiring. Eventually, you are asked to fill out routine personnel application forms which are held for future reference. Update these forms every 6 months.

AUDITIONS

If there is an opening at present or in the near future, you will be asked to audition. An audition provides the station management with something concrete to look at and listen to and pass around to others for an opinion. An audition usually shows whether a person's appearance, movement, and voice have sufficient impact on the management, since few hiring decisions are made by one person. You should be less concerned about reading fluently (which can be improved in rehearsal) than with radiating a confident, warm, unique personality—that is, something marketable. Frequently, you will be asked to read copy without preparation. In fact, some news directors want to know whether a person is fluent at sight reading. This tells a lot about how the performer would sound in an ad-lib situation, potential errors that exist in voice, or judgment, and whether there is a keen ability at word recognition. Generally, professional employees do not want to train a nonprofessional, for stations are not in the business of training people. They will acquaint an individual with the specific way they handle material and with their equipment, but they do not have time to train a novice. So, the audition is to show what you can do as a professional.

An audition can be nerve-racking, but you should keep in mind that you probably would not get one if there wasn't some interest in your ability. Even a short audition takes staff time and equipment. Management reviews auditions only of professional media performers. It is not uncommon for 40 or more professionals to audition for a news anchor position in a major market. Hundreds audition for parts in films and other media jobs. Typically, auditions are conducted selectively for those professionals who appear to be most qualified (they have established ratings in their markets) and may have an agent.

Performers in commercials are usually cast through a casting department maintained by a local advertising agency or through an independent casting firm. Some photographers may serve as casting directors from time to time. You should be acquainted with local casting agencies. If you get a *call back* after an initial audition, you should try to perform the same way the second time, because apparently your work was acceptable. Numerous auditions are often held to cast one spot. Regardless, preserve an air of confidence.

RÉSUMÉS

Nearly every employer appreciates an in-hand reminder of the talent. A single photograph or *head shot* showing a close-up of the person and giving minimum vita (name, telephone answering service, hair color, weight, age) is a common way of doing this. The pose should be appropriate for the job you are seeking, and the photograph should be made by a professional photographer who is in the business of taking head shots. The close-up should show what you actually look like, including marks and lines of character in the face. Glossies are usually preferred over matte photographs. Some casting agencies prefer 2 x 2-inch color photographs, but 8 x 10-inch, black and white glossies are typical. A *composite* includes various photographs, principal credits, and vitae. A composite is usually for models. Do not submit a composite for a reporter's job. Keep a *book* or collection of your best work including articles and photographs. Assemble your credentials attractively. They document the experience you have had. A little showmanship is desirable, but directness, good pictures, and credits are what employers are looking for. A sample résumé is shown in Figure 6.

Videotapes

A brief videotape can be made at a professional recording studio. Some schools have such recording studios, and can do a respectable job. The performer is usually shot close up most of the time, with medium or long shots included as required. The performer presents 5 minutes of copy in a sampler format, with most of the time de-

YOUR NAME
YOUR ADDRESS
(— — —)

Objective:
To pursue a career in news broadcasting and to eventually become a news anchor.

Employment History:

Current — CNN, Atlanta, GA

News Intern
Gather facts and information, assist in production, and prepare audio tapes for programs.

June 1989 to December 1989 — KRIV-FOX, Houston, TX

News Intern
Responsibilities included working on the assignments desk, field reporting, gathering information, and editing.

January 1989 to June 1989 — Major Industries, Inc., Houston, TX

Graphics Artist
Designed various presentation packages using different software programs. Wrote and edited material in packages.

Education:
University of Houston
Bachelors of Arts in Radio/Television/Journalism
Graduated with honors, May 1990

Awards:
ABC Press Scholarship

General Information:
Fluent in German.
Proficient in basic computer skills, graphics, word processor. Member, Volunteers for the Red Cross.

References:
References, audio and video tapes available.

(A photograph accompanying this résumé is recommended.)

▶ ***Figure 6*** *Sample Résumé for News and Information Positions.*

voted to the principal work for which the person is interviewing. Videotapes can be very expensive and should be sent, in duplicate only, to stations selectively with return postage guaranteed. Needless to say, everything about the tape should be perfect. Remember that the tape is designed to showcase the talent and not to exhibit the cleverness of the director, although imagination in selecting flattering shots and other directorial approaches should be encouraged. A brief tape might include a news report, an interview, and an air check.

Credits

Employers always ask applicants, "What work have you done, and where have you done it?" You should have a list of all previous experience, especially on-air commercial jobs and films. These are *credits*. You want to be associated with top stations, film companies, networks and recording studios, and well-known people in the business. This association aids in establishing your reputation. Your record should also show that you progress to better stations, bigger markets, more responsibility, more money, and more benefits. Some beginners, listing their credits on their employment forms, are surprised to learn that experience is more important to an employer than education, at least initially. You should expect this. Paid experience indicates that you are a professional performer progressing in a highly competitive business, that you are aggressive enough to get a job, and that some employer was willing to pay for your services. Each credit is listed with great care, especially by a novice who may not have many. As years go by, the principal credits are listed until, of course, the performer is so well-known that the networks, stations, and film companies feel fortunate to get a contract with the performer no matter what the price.

PROFESSIONAL ASSISTANCE

If you become successful, you may gather quite a collection of other professionals to assist you. The first is likely to be one or more instructors to develop your talent; the next, an agent to get more work; the third, a business manager to invest your money; the fourth, a lawyer to draw up contracts; the fifth, a companion or secretary to handle correspondence, personal maintenance, and, in some cases, protection; and the rest constitute a production staff. A major program may require 100 people. One half-hour weekly series on a local PBS station has about six people: producer-hostess-writer, director-editor, production assistant, secretary, part-time camera operator, and business and legal assistants.

Schools

There are schools for everything. Schools are designed to keep performers mentally and physically fit through discipline. Schools keep performers in practice, which is especially useful when the performer is not working constantly. Schools give a performer a chance to perfect skills—voice, appearance, movement—and to bring these skills to the same degree of accomplishment as the competition provided by other performers. Schools enable a performer to become acquainted with other performers so that he or she can draw upon other performers for ideas, inspiration, and comparison. Schools often help performers find work, especially entry-level jobs, before they are experienced enough to attract an agent. Schools can be very expensive and time consuming; therefore, you want to be certain the time and money will serve to promote your career objectives.

While some professional schools may be able to train commercial announcers and others in specific vocational roles, most news and information specialists need a much broader education. Colleges and universities aid the development of a performer's talent within the context of a liberal arts education. Even there, concentrated studies in communication—radio, television, film, and journalism—must be supplemented by traditional disciplines such as languages, history, political science, arts, sciences, and mathematics. Two decades ago, on-job-training at wire services and newspapers was adequate; nowadays a news and information specialist really needs a college degree.

None of these schools should be confused with those few that claim shortcuts to success. There are no shortcuts at any price. A performer should examine the reputation of the school before entering it. Ask for a list of graduates, or get the opinion of professional performers in the market. All schools do not have equal reputations for turning out people who are fully prepared. A school depends on its faculty and its graduates for reputation, and so should the performer. The performer should also get biographies of the faculty, and should see the facilities. A school contributes to the foundation of a performer's career, and so a proper choice is extremely important in the long run.

Instructors

An instructor is not necessarily a performer, but instructors know what it takes to be a performer. They prefer searching for new talent. Some instructors may not have the appearance, voice, or other characteristics to be a performer; instead, they may have the patience, concern, understanding, comparative knowledge, and intuitive sense for developing talent. A good coach is very important to a performer. Besides discovering those with the greatest potential, instructors encourage performers, with the care of a gardener nurturing tender plants—giving discipline, love, and attention when they need it, and finding opportunities for them. Your instructor may be the only source of human hope and inspiration you have through much of your career, knowing that regardless of success or failure, the teacher is the one understanding person who knows the pitfalls you endure. Many fine teachers are found in high schools, colleges, universities, and professional schools. You may meet others on the job. You will indeed be fortunate to continue your education under the supervision of a highly respected news director, TV or film director, talent agent, producer, or other mentor.

Talent Agents

After you are sufficiently well-trained to be placed for continuous work, you are ready for a talent agent. Sometimes talent agencies take performers who show promise at an earlier stage; however, agents are essentially looking for persons whose abilities they can sell. A talent agent is a salesperson whose chief function is getting the performer jobs at the highest salary the market will bear, and for this service the talent agent gets a percentage of the payment negotiated for the talent, commonly 10% of the gross. Thus, the more money the talent makes, the more the agent gets. The benefits of a talent agent to a performer are (1) an intimate knowledge of what is happening in the industry; (2) an optimum presentation of the performer's assets to buyers; (3) ongoing contacts with talent buyers; (4) a detailed negotiation of the contract; (5) advice about career development, and, in some cases, the security of a friendly counselor. The principal drawbacks are that (1) the agent may have to divide attention among many clients; (2) the agent may receive a commission from work the performer obtained alone; (3) an agent may tend to favor the position of major buyers in negotiations; and (4) the agent's advice may be bad, and the agent may benefit more from the association than the performer, in which case the performer should get a different agent.

Local talent agencies provide performers with job leads in their immediate vicinity. Agents in New York and Los Angeles provide talent for major markets and often for programs distributed nationally. Many celebrities in the business, whether in journalism, the talk circuit, or sports, have agents. Generally, performers do not talk to anyone seeking their services; instead, they refer potential employers to their agents. The agent determines whether the performer should do the work and whether there is enough money, prestige, or both in the job. The talent receives a secondary benefit in this arrangement. Any blame for being unwilling to meet a request is placed on the agent, and not on the performer. An agent schedules the performer's commitments, making certain that proper decorum and amenities such as accommodations, receptions, and billing are met in accordance with the contract. A large talent agency is active in every type of media performance for national networks and film companies. These agencies represent individual performers or they may package entire programs. There are four ways as a performer that you may come to the attention of a major agent: (1) be referred by a mutual acquaintance, (2) be performing where an agent can see your work, such as on the air or charity benefits, (3) be an established performer, (4) be aggressive enough to take your résumé and photographs directly to the agency, and ask for an appointment with an agent. From time to time talent agents review these materials and set up appointments with performers who look promising. Meanwhile, there are a number of smaller agencies that may meet your needs as well as or better than a large agency. Talent agents prefer 3-year contracts, but performers may negotiate for shorter periods and with more than one agent, because many agents specialize in different media. Be sure to read your contracts very carefully.

Managers

A performer may have need for a personal manager, business manager, or lawyer. A personal manager performs tasks similar to those of an agent, but having few clients, he gives a performer custom service. The combination of an agent and personal manager may amount to 25 % of the performer's salary; this expense should be more than offset by performer revenues. Business managers and lawyers (they may be the same) specialize in handling a performer's money. If you make a lot of money, you should invest it wisely. There may come a time when you are out of work, with nothing to live on for months or even years except investments that were made when you were doing well. Many performers are wealthy because of the money they invested with the help of managers in real estate, stocks, and other businesses. Business management has a risk of course, and you should select a completely reliable manager, who will make certain your investments are accounted for. While performers appear to make vast sums of money, frequently they spend equal amounts. For contracts involving much money or complex commitments over a long period of time, a performer or the talent agent is advised to pay for the services of a lawyer specializing in talent contracts.

Personal Staff

Performing can be a lonely life, especially for speakers who are on the road as well as appearing in media; and many performers, particularly women, take along a paid companion or secretary. The secretary may handle appointments, correspondence, and the performer's personal upkeep; but she is usually much more, giving friendly advice and encouragement when needed. Occasionally even a local performer—say, an investigative reporter who is threatened or a reporter at the scene of a riot where some have been stabbed or killed—may have to have protection. Celebrities at all levels quickly become wary of public demands as their prominence increases.

The Production Company

A media celebrity often forms a small production company for the express purpose of producing the programs in the series. This unit, an independent business, has experts who collectively produce the vehicle in which the performer regularly appears. In fact, the resulting videotape series may then be owned not by the network, but by the performer's company, which syndicates the videotapes for years to come with a profit continually flowing to the performer and financial backers. Once the realm of the rich and famous, numerous others, as already cited in instruction, ministries, and feature magazines, package an entire series for broadcasting and cassette markets that the company owns and leases.

LEGAL AFFAIRS

Performers should be aware of the provisions in most contracts, so they can balance the desirability of these provisions with the chances of getting a job. They must be aware of certain laws pertaining to the limitations of program content, and the conduct of a performer on-the-air.

Contracts

A performer may become involved in several kinds of contracts. Commonly, one is for performer services. But, in addition, other contracts may be written for scripts, musical compositions, performer endorsements for advertising or promotion, and for the development of audio or videotapes, films, or other media entities that may be distributed for profit. If the performer has an agent, the agent usually handles the contract, because it is in the agent's best interest for the talent to be well taken care of. Some performers, like newscasters who may not have agents, may want a lawyer to go over an individual contract with a television station. You may wish to have your own lawyer read the contract. Some of the provisions of a contract include: (1) basic salary and increments, (2) amount of work to be done on and off the air, (3) compensation for personal appearances, (4) time off for vacations and holidays, (5) fringe benefits such as insurance, retirement, vehicle expense, and office accommodations, (6) release time to do other projects, (7) billing and credits on the program(s), (8) use of the performer's image or voice for endorsements or promotions, (9) residuals, if any, (10) expense account, (11) staff assistance, (12) and period of the contract with provisions for early termination or renewal.

Restrictions and Ethics

While your contract may be your initial concern, once you gain a measure of prominence, you will sense the care with which you must conduct your public, and often your private, life. Performers and the companies they work for may, justifiably or not, become the targets of litigation. Whether in information or entertainment, the basis of the litigation is usually fairness, or the perception of fairness when bringing to public attention the actions of individuals and companies. Because of the enormous power of television, a performer can inadvertently ruin someone else, damage a company, or destroy a cause. Anchors, contributing reporters, commentators, moderators, panelists, analysts, and talk show hosts must aspire to the highest levels of presenting material fairly. In accounting the day's events, in assigning meaning to them, in allowing an exchange of views and criticism, the performer must check the accuracy of data, attribute statements, divide fact from allegation, and allow reasonable presentation of all points of view, even those he or she may not share. Without such openness, the performer subjects himself and the company to possible litigation, severe criticism, and public disfavor.

You will find it worthwhile to consult federal and state laws concerning the prohibition of lottery, fraud, obscene language, libel and slander, and the assurance of fairness in contests and in the treatment of political candidates and public figures. While companies retain attorneys to guide you in these matters, most problems will never occur if you retain high ethical standards in your judgment of individuals and situations.

SALARIES

Your salary is based on the demand for your services. As a novice, your compensation may range from unpaid exposure to a few dollars an hour. Ten to 50 dollars for a single appearance is not uncommon. A beginner prizes opportunity even more than money. A great deal depends upon how much money is available for the job, and how highly the employer rates your work. One novice sportscaster paid his own expenses to travel with and learn from a network crew on location. Often a second job pays the bills. If possible, get a related second job, but remember you are taking it mainly for the money. It is a means to an end. A college graduate who enters the information business should be able to subsist on the company's salary. You are usually so busy that you do not have time for a second job. Because of the hazards of starting out in a large market, you are far better off if you can go to that city with some money in the bank or financial support from your family.

Estimates of salaries for radio and television performers are difficult to predict due to variations in station profits, market size, and requirements of the job. Visibility in the community and on the air pays the most. Of course, station management will attempt to get talent for the lowest possible figure, while the talent or its agent will try to get as much as the market will endure. Typically, an entry salary in news ranges from \$10,000 to \$12,000 a year. Once established, in about 2 years it jumps to the mid-teens, and later the twenties and thirties. Anchors get more money than rank and file reporters: $25,000 to over $100,000 per year. Company benefits may add another 20% to your salary. While a comfortable living can be made in news and information programming, the much publicized multimillion dollar contracts are rare.

RESIDUALS

Sometimes a performer is very successful in a single role, and may or may not be successful in anything else. This performance may be in a media series or a commercial. The initial payment to the performer is followed by the continuous payment of smaller amounts every time the TV series, spot, or film is shown. This is known as a *residual*. You may receive payment for a number of years from a television program or series that ran only briefly.

UNIONS

Network operations and major film companies—those businesses which make a lot of money and whose programming has wide distribution—are usually unionized. Some exceptions may be found in states like Texas that have right-to-work laws. Houston, the nation's fourth largest city, is the largest non-union market. Most broadcast stations are nonunion. Non-union performers get whatever compensation the market will bear; union performers get whatever the union contract agrees to or more. Generally, to enter a union, a performer has to be employed in a job covered by the union contracts. Some performers belong to more than one union. The American Federation of Television and Radio Artists (AFTRA) and the Screen Actors Guild (SAG) represent media performers. Contracts include all work whether live, film, tape, TV, or radio. Actor's Equity Association (AEA) represents those who appear live on the legitimate stage, in stock companies and industrial shows; and the American Guild of Variety Artists (AGVA) covers nightclub performers. The American Federation of Musicians (AFM) represents musicians, and the Writers Guild of America (WGA) has jurisdiction over writer contracts in its two independent chapters WGA-East and WGA-West.

AFTRA includes information specialists, such as news presenters, reporters, and sportscasters; entertainers, such as singers, dancers, and animal acts; and commercial announcers. SAG, which was a theatrical motion picture performers' union, retains jurisdiction in that category, and represents performers in all TV shows and commercials made on film. AFTRA and SAG have established minimum wages, called *scale*, with minimum rehearsal days and hours for a day and for a week. Wages are tied to the length of the show and the use—local, regional, or national. There are various scales depending upon the number of lines the performer has. For example, principal performers, those with under five lines, extras, voice-overs, and so forth. One newscaster frequently moonlights—sometimes in the role of a newscaster—in feature films.

AFTRA and SAG contracts are complicated and long, in part because they address themselves to various categories of performers, such as those in news, commercials, or drama. Essentially, contracts concern payment to and working conditions for performers. A player (performer) is paid specified fees for each session. SAG has per day fees, but AFTRA works on a per program basis. Player compensation is determined by whether the player is on- or off-camera; whether appearing in a large or small market; whether the duration of service requires overtime, overnight, weekends, or holidays; or whether the program is distributed nationally or internationally, in which case residuals are paid. A player may be compensated for use of his or her own make-up, hairdressing, street or evening clothes, and travel time.

A player may expect certain working conditions such as adequate dressing rooms, first-class travel accommodations, rest periods, and protection from pay loss due to injury on the job. The contracts may provide protection clauses concerning dangerous or hazardous circumstances, foreign duty, and conditions for living overseas. Union contracts include non-discrimination clauses that encourage the use of women and minorities. Penalties to producers and players are defined if the conditions of the contract are not met.

11

Instant Basics

Unexpectedly, you are asked to be on television. Most likely you are being interviewed or in some cases you are appearing as a panelist. You are a physician, attorney, educator, business executive, politician, minister, social worker, criminal, victim, eyewitness—potentially everyone could be on television as a performer at least for a few seconds. An invitation to be on television is exciting, and you want to do your best. You are probably familiar with typical television presentations. You have heard hundreds of interviews and reports, and you have seen hundreds of talk programs. You have a good idea of what a television studio is like, even though you may never have actually been in one. The circumstances are therefore somewhat familiar. Nevertheless the experience is new to you, and you are excited, if not anxious, about it. After all, untold numbers of people see the program, and you want to be acceptable to yourself, your friends, and your audience. The following are the principal items to keep in mind.

EMPHASIZE CONTENT

You are asked to appear on television probably because you know something or do something unusually well. The public is interested in learning something new, or watching a person who has perfected an ability. So never forget why you are on television. When the interview or act is over, be sure that you have said or done whatever you came to do or say. Focus your remarks on your knowledge or skill, so the public has a clear understanding of who you are and what you do. Discuss what you know, avoid what you don't, and do not speculate. If you are a demonstrator, keep your appearance brief and of high quality. The length of a typical appearance is from 20 seconds to 3 minutes. Five minutes is considered a long time, even for a professional performer, wherein the professional performer is providing the content. Exceptions may be found in noncommercial programming. For the most part, the content should be the essence of what you do.

ORGANIZE YOUR MATERIAL

Time on television must be carefully spent. If you are well prepared, it will go by quickly, but you will have made your point. Think through what you intend to say, and ways to say it in the fewest words. Your comments should be concise, preferably in short, easy-to-understand sentences. If you take time to ponder on camera, the conversation will seem slow and dull. Make your point in a short expression. Don't expect to ramble or fill time. Interviewers who want to know more should ask additional questions, one at a time. You can anticipate the questions, rehearse your phrasing, and have tentative answers based on what you think the public must or wants to know. Don't sound mechanical, of course, but do sound precise, knowledgeable, and articulate.

Besides verbal preparation, performers should have their visuals fully prepared. Visuals include studio cards, drawings, charts, photographs, books, art objects, animals, machines, and a long list of arts and crafts. In addition to three-dimensional objects, slides, videotapes, and films are commonly used as visual reenforcement. Be sure these items are in the proper aspect ratio (4 units horizontally to 3 units vertically), are properly mounted for presentation, are easy to recognize by the viewers, and are arranged in the order of use on the program. If you are to show these items, hold them steadily in one place, in view of the proper camera lens for about 5 seconds. Frequently, visuals are shown in detachment or *limbo*; that is, in a separate area. Keep track of the visuals before, during, and after the program, so that items are not lost.

ENJOY YOURSELF

Have a good time with the television experience. Very few topics are grim. Most subjects, and acts, even if briefly sad, are light and pleasurable, so keep them that way. The public should enjoy them to the utmost. For it to do so, you must enjoy telling about the subject or presenting your work. Energy, your smile, and appropriate laughter are enormous aids in presenting the material favorably. When you are in some other person's presence, when you are in someone else's home as you are on television, you are there mainly to be pleasant, to have a good time, and to encourage that viewer to invite you again. When you are on television, conduct yourself in much the same manner as you would if you were visiting someone else's home in person. Even though you are nervous, and most performers are nervous, concentrate on the value of your information or the ability demonstrated in your work. Have fun in your interaction with the host or others on the program, and be cooperative and friendly with the television crew. If you are having a good time, the host and the studio crew will have a good time, and so will your audience. Even if something goes wrong, avoid being upset. Instead, smile your way, undisturbed, through it.

LISTEN

There are two aspects to listening. One is listening to the instructions that will make the program run smoothly, and the other is listening to the host and participants during the program so that you respond to what is being said on the air. The producer and host may give you directions before the program. These instructions will have to do with the content of the program: how the topic will be covered, where the breaks will come for announcements, and how the program opens and closes. The production crew may tell you where to sit, how loudly to speak (which is usually loud enough so the crew can hear you), and often adjusts lighting and make-up when you are on the set. Let the crew decide where to position cameras, microphones, scenery, chairs, and lights. Typically the host, producer, or floor director will give you specific instructions. If none are given, try to act normally. Do not look in the camera lens unless you are told to. Instead, look at your host, or if acting, beyond the camera lens. Don't worry about whether the program is going well or whether the crew is doing a proper job. You can't do much about it. Do not touch the microphones or move the furniture, even though you may feel uncomfortable. They are placed to enhance visibility for the camera operators and to avoid shadows. Concentrate on doing your part correctly.

To be effective as an interviewer or interviewee, you must listen carefully during the program. Often an interviewer must hear instructions over a headset and pay attention to the guest simultaneously. In order to have a meaningful dialogue, the host and guest(s) must listen to each other and develop a conversation from the most interesting aspects of what they hear.

KEEP YOUR COMPOSURE

Television can be tense. Everyone is trying to do well. Circumstances in some newsmaker situations and controversy may be unpleasant. Even if your interviewer, other guests, or climate is hostile, you keep your composure by being polite, warm, and friendly. Often as newsmakers, performers find themselves in difficult circumstances, under duress, or debating complex, no-win issues. Nevertheless remain calm. Even if you are caught in a circumstance that requires you to make "no comment," retain a manner that is confident, cordial, and understanding of the job the press must do. Avoid getting excited or provoked, because you may say something you'll regret. Do not answer questions you have not considered previously. Remember all videotape and film that is taken of your appearance will be available for others to rerun forever. You cannot retract a statement effectively once it is recorded. Politicians are especially vulnerable when under stress. They have done things and said things that haunt them the rest of their lives. Alleged criminals have made confessions which later they recanted on advice of counsel; but the statement is never really erased in the public's mind. Most of the time you will be trying to impress the public with your charm, energy, ability, and knowledge, and you can do this most favorably by being consistently warm, friendly, and generally happy.

PRESERVE YOUR IMAGE

Image is everything. The way you *seem* to the public is the way you *are* to your viewers. Those who are watching have little to go on when assessing an individual. What they see or read is probably the extent of what they know about a performer's work. The public quickly assesses you and your work. First impressions are enormously important. The public's perception is based largely on television, because it is the most influential and easily understood medium. Television doesn't reveal everything, however; it only shows what is presented to it. For instance, visually speaking, television has a filtering effect. Because of electronic scanning, performers are well defined but not highly or minutely defined; hence, a performer tends to look more attractive on television. To wit, all of the facial and other imperfections do not show. Then too, sheer brevity on the air gives the public short perceptions of an individual. A performer or producer's ability to edit tapes within a fraction of a second enables a performer's work to reach a high level of perfection. Thus, you should do everything you can to make certain whatever is presented on television is the image you want projected. This is not always easy to do, for television is basically a group effort, and your best efforts may be jeopardized by someone else. The result is that if you have faith in your producer/director, you will probably be pleased with the outcome. That particular view the public sees of you and your work constitutes your image.

DRESS CONSERVATIVELY

The most acceptable attire for television, especially for news and information, is conservative. For men, that means a medium to dark suit of contemporary design, a pastel shirt, a low contrast tie, few accessories, hose that match the trousers, and polished shoes. For women, that means a dress or perhaps pant suit, preferably of one color and simple design, hose, high-heeled shoes, and little jewelry. A skirt with some fullness is usually more comfortable than a tight skirt. Clothes should fit well, being neither too snug nor loose. With a proper emphasis on content and your personality, fussy clothes, sparkling earrings, shiny watches, bold plaids, or accessories that carry messages through symbols are unnecessary and unwanted. Avoid high contrast clothing, for it makes electronic adjustments in picture quality difficult for engineers. A dark complexion next to a white shirt may result in a substantial loss of facial definition when the engineers adjust for signal requirements. Select clothing that favors the medium range of colors; that is, the colors are evident, but the contrast is minimal. Theatrical or special costumes may be desirable if they are appropriate to the activity, such as athletic gear for sports or formal attire for fashion shows and fundraisers. Outlandish costumes rarely have their place, and should be coordinated with the producer of the program. If in doubt, a performer should bring along a second outfit to the studio, and let the producer/director decide which one is better for the occasion.

WEAR BASIC MAKE-UP

Make-up is used to enhance a persons definition and to give energy to your look. If you are Caucasian, you will probably need some make-up, unless you have clear skin that is tanned. The exception is the pale white woman who wants to preserve that "snow white" appearance. Brown complexions need little make-up; blacks require good lighting to assist in aiding facial definition. The make-up sufficient for most men and women is foundation that conceals blemishes and yet maintains a healthy look under bleaching hot, white studio lights. At the very least, a dry pancake or powder will absorb perspiration that is bound to appear on faces, foreheads, and tops of bald pates. You should have your own make-up applicator, and a small hand mirror for a last minute touch-up. Men do themselves a disservice if they think they are too masculine to bother with make-up that would cover blemishes, beards, and an energyless complexion. Two big problems for men are beards that make them look like derelicts and bald heads that resemble mirrors. Additional make-up is necessary if you want to give your face further dimension, alteration, or decoration. Eye shadow, eyeliner, and eyebrow pencil are typical aids. Women and men can profit from a limited use of these same items so long as they do not appear artificial or effeminate. Error on the side of too little rather than too much make-up.

BE DEPENDABLE AND ON TIME

It goes without saying that you should be recognized for your talent, but you are also judged on your reliability. Don't keep everyone waiting for you to show up. In business, knowing a performer can be counted on is almost as important as being highly touted by viewers. Many performers keep their jobs because they are capable, competent, and absolutely dependable. Many television programs are routine in structure; they do not require extraordinary ability. Yet, such a series may be very popular and a big moneymaker. Meeting a tight, economically viable schedule, and meeting the demands of the job are highly prized assets in a performer. Showing up on time, ready for work, knowing the material, being there day after day or whenever called upon, are attributes that, in their way, are just as important and valuable as award-winning ability. The level of dependability you establish with a producer or company may guarantee requests for return appearances. Television programs are often produced by a small group of persons who trust each other and know each others' talents. This group forms a unit, and the members of the unit tend not to let each other down. They all work for the benefit of the program series. If you are invited to participate, it is because someone in that small group believes in you. Whether a program originates at a local station in a small town or at a major network production center, performers must be dependable.

RETAIN CONTROL

Performers who appear irregularly on television usually look satisfactory, are well thought of for a moment or two, and are quickly forgotten. This is the role of the amateur in media. A professional performer in media, on the other hand, wants to make a positive appearance that is remembered and recycled for a long time, such as in a television series. Because of the potential for longevity, you want to take control by having the series refined and rerun indefinitely. When you become a celebrity, wealthy, and own the show, you will have great control over the product or program series, and you will reap the financial benefits as well. Top performers reach a certain lofty level where no one else has gone before and, consequently, they must decide what is best for their future, win or lose.

A beginning performer retains control through being sufficiently prepared to do a good job in the first place. The key to doing a good job is having a script or outline. Very little in television is ad-lib, even though it may seem to be. Insofar as possible, get your producer/director and crew to work out a script or plan for what you are going to do. Be politely aggressive by making suggestions, but at the same time sense when to yield to others who are in charge, who probably have more experience than you do, and who, in the long run, want to make you look attractive.

Keep in mind that you don't have to appear if the situation isn't right for you. You don't have to talk to reporters or let them in your home, and you don't have to go to the studio or talk to reporters on the phone. Remember you have those options.

Now do your best with high hopes for success. Luck plays a part. Whatever happens, do not be hasty in making judgments or drawing conclusions. Give yourself a chance to gain experience and to mature. Don't expect instant feedback and recognition. If you remain in television and prepare thoroughly, your opportunity will come. When it does, be ready to take advantage of it.

Glossary

Act A performance. A display of a performer's ability.

Action A verbal cue from the director to begin performing. Often extras are cued initially to create atmosphere, hence the cue "background action."

Actor/actress A performer who appears in dramatic roles portraying someone other than himself or herself.

Actuality The voices or sounds of an event at the scene.

Ad-lib Spoken remarks that are not written in the script.

AFTRA A talent union called the American Federation of Television and Radio Artists. Referred to as "Af-tra."

Air check An excerpt from a broadcast illustrating how a performer looks and sounds on the air.

Amplifier An electronic device for increasing volume.

Announcer Various performers who present information, news, commercials, sports, and weather. Narrowly defined, an announcer reads commercials, promotional spots, and station continuity.

Audition A selection and testing process wherein producers and directors determine which performers to hire.

Back timing The process of predetermining where a performer should be near the end of a segment or program so, with careful judgment, the program will come out on time.

Background A term, abbreviated "B.G.," that refers to maintaining music or sounds, or both, at a low enough volume so that the performer can be heard.

Bidirectional A microphone pickup pattern resembling a figure eight that allows two (or more) performers to speak opposite each other while using the same microphone.

Bleep An electronic interruption in verbal continuity, usually deleting a word or phrase considered in bad taste or legally injurious.

Blocking The plotting of action by the director during rehearsals.

Book A multi-purpose term referring (1) to the periodic listing of program ratings, (2) a script for a musical show, (3) a performer's collection of credits and photographs. *To book* means to schedule a performer's act.

Break To stop action, i.e., to take a break. Also, to move talent and cameras to another set.

Breakaway Scenery or properties designed to break on cue.

Build The gradual development of interest or tension in a segment to the point of climax.

Business Bits of characteristic action like smoking, knitting, or shaking used by an actor to portray a role.
Bust To stop or terminate action.

Call A warning to performers for the beginning of a rehearsal or on-air session. Also, a notice of a need for performers; i.e., a casting call.
Cameo A brief appearance in a program.
Canned A prerecorded performance; i.e., one on tape or in a film *can*.
Cardioid A heart-shaped microphone pick-up pattern.
Cartridge A small plastic box containing a tape loop that can be recued for immediate playback.
Cassette A small reel-to-reel audio or videotape contained in a plastic box that is capable of playing an entire program.
Cattle call A call for numerous extras.
Charisma The ability of a performer to consciously or unconsciously exude a compelling personality which allows him or her to retain an audience.
Chroma key A production technique which electronically inserts a performer into a television picture. Under these conditions, the performer should not wear blue.
Character A performer portraying an individual he or she may not resemble because of age or personality.
Cheating An effort by a performer to create the illusion of talking directly to someone (e.g., an interview), while positioning the body for a maximum open relationship with the camera.
Circular response The delivery of a message from a performer to the audience and the return of the audience's response to the performer.
Climax The highest point of tension or interest in a performance.
Close The conclusion of a program. *A closed position* is the relationship of a performer's body to a camera that allows the least expression.
Close Up (CU) A close-up shot, usually the face.
Clothing Regular clothes worn on camera.
Cold The reading of material without prior knowledge or rehearsal of it.
Communication The ability of one person to evoke a response and/or create an impression on another person.
Composite A collection of photographs illustrating the versatility of a performer.
Concentration Undivided attention.
Condenser A high fidelity microphone.
Conflict Dramatic opposition of the protagonist with one's self, another human being, society, or fate.
Control room A room adjacent to the studio. It contains audio and video consoles, which are utilized by the director and engineers during production.
Credits A list of work experiences including all live and media performances. Also, a list of names and titles at the beginning and/or end of a program.
Critique An evaluation of the strengths and weaknesses of a performer.
Cross A movement from one place in the studio to another.
Cue A signal for a performer to begin the action.
Cut A signal to stop action.

Cyclorama A huge drape or plaster (*hard cyc*) wall providing the backdrop for a setting in a studio.

Diaphragm A large muscle located between the lungs and the abdomen. It serves as a bellows in the breathing process. Also, a flexible element in a microphone producing electronic impulses.
Direct contact Eye-to-eye contact with a live audience and/or by means of looking directly into a camera lens.
Directional The dominant pickup pattern of a microphone; i.e., uni-, bi-, omnidirectional or cardioid.
Director The principal person responsible for the artistic development of a production.
Dry run A rehearsal which is not recorded or filmed.

Electronic Field Production (EFP) Segments shot live or on tape at the scene. ENG refers specifically to electronic news gathering.
Entrance A performer's movement into view or recognition on microphone.
Exclusive contract A legal agreement that binds a performer to one company or agent.
Exit A performer's movement out of view or microphone pickup.
Extra A performer without lines who contributes to the background, atmosphere, or crowd scenes.

Fact sheet An outline and/or list of titles or topics to be discussed on a program. A background sheet or biography of a guest. Sometimes called a *run down*.
Featured player A performer requiring special attention, although not a star.
Feed a message sent from one performer to another, usually over a headset. Also, a news summary such as an electronic feed.
Feedback A performer's response from the public, critics, and those associated with the production.
Freeze To maintain an exact position.
Format The overall structure of a program or series. A type of script listing major programming steps, including usually a scripted open and close.
From the top An expression meaning to start at the beginning.

Gobo A sound absorbing screen used to isolate performers or instruments.

Hand signals A silent method of sending instructions to performers while they are on the air.
Hand props Personal properties carried by a performer.
Hard news A news story about an event having significant current impact usually on many people. It often deals with disaster, economics, and political events.
Head shot A close-up of the face.
Hold To stop all movement in action or speech, usually for laughter or applause.

Idiot card Large cards with lines of dialogue or cues that are hand held in easy view of the performer.

Key light The brightest major light source for a performer.
Key map A locator map for news departments that send reporters to the scene.
Kicker An amusing story at the end of a newscast.
Kill To cut, remove or get rid of action, lines, stories, or production elements.

Lavaliere A tiny microphone held around the neck by a lanyard or clip.
Leading man/woman A performer having the major responsibility or role in a program, such as a documentary.
Level The volume of a voice as registered on a volume unit's meter on an audio console. "Give me a level" is an engineer's request for the performer to keep talking until audio equipment is adjusted properly.
Lines The dialogue a performer must deliver.
Live A performance in which the talent is working in the presence of an audience. Also, the direct transmission of a performance at time of origin.
Long Shot (LS) A long shot, a full length view of the performer.
Looks A direction in which the performer will be asked to look in order to match a previous shot, to prepare the audience for action, or to indicate the location of a person or object in the shot.

Marks Designations usually taped or painted on a studio floor indicating positions for performers.
Master The original recording of a program, disc, or film.
Medium Shot (MS) a medium shot; that is, a view of the performer from the waist up.
Message A communication between a performer and other persons, animals, nature, and certain objects.
Mixer An electronic device, usually an audio console, used to combine sounds.
Monitor A television receiver without sound placed on a studio floor, news desk, or in a control room. Also, any of several audio receivers found in TV, radio, or recording studios.
Morgue A reference file of stories, photographs, films, and tapes.
Motivation The purpose or reason behind a performer's movement.

Nielsen A term referring to a program rating, probably measured by the A.C. Nielsen Company, in which the performer appeared.

On To be on the air, camera or microphone.
One sheet A photograph with minimum vita, used for résumés.
Open The beginning of a program. Also, the relationship of the body of a performer to a camera that allows fullest expression, *an open position.* A live or *open* microphone. *To open up* means to be freer in movement and interpretation.
Out-cue Those words constituting the final seconds of a segment, such as the final seconds of a sound bite. An out-cue is critically timed to match the upcoming segment.
Overplay To overemphasize dialogue or movement.

Pace The overall rate or flow of continuity in a production.

Pad A period of flexibility in the continuity that can be expanded or cut near the end of a segment so that it comes out on time.

Pager A beeper device which is activated by a radio signal to call reporters.

Pancake A water soluble make-up foundation.

Pan stick A foundation in stick form used to cover difficult areas such as beards.

Pantomime The silent expression of an idea or emotion through body movement.

Performer Anyone who appears on camera and/or microphone, especially those who are paid to do so (media professionals) and those who do so because of their expertise in areas outside media (community experts). Narrowly defined, a person paid to appear on television.

Personality The innate charm or essence of an individual. Also, a performer whose personal characteristics dominate whatever role he or she portrays is referred to as a *personality*.

Phonation The formation of a sound by means of the vocal mechanism, dominantly the lips, tongue, teeth, and surrounding areas.

Pickup The range within which a microphone can get a sound signal. A *pickup pattern* refers to the shape of the sensitive area surrounding a microphone. Also, to accelerate the pace of a program largely by means of responding more quickly to cues; to begin where the action last stopped.

Pitch The range of vocal sounds a performer can produce from deep tones with relaxed vocal folds through falsetto.

Playback The replay of a performance on audio- or videotape.

Player A performer; a common term used in union contracts.

Post synchronization A process in film production in which the sound is added after the picture has been recorded.

Presence The optimum clarity of sounds. Also, the psychological status of being in the same room; i.e., present.

Presenter Any performer appearing live or in media, usually as a news or information specialist.

Producer The executive who arranges the business aspects of a production, including the contracting of a director and major performers.

Promo A promotional announcement or reference.

Prompter A human or electronic system of cuing talent.

Prop An abbreviation for property. All objects necessary for the completion of a setting and for use by performers. Usually this includes furniture and smaller personal items; that is, *hand props*.

Publicity Attention which media, especially news, gives to a performer without charge.

Quick study A performer who can remember lines easily.

Range The extent to which a performer can produce high to low pitch levels. Also, the extent of one's ability to perform in various capacities.

Ratings An index of the relative number of persons tuned into a radio or television program.

Resonance The reenforcement of a vocally produced sound by means of physical structure, acoustics, and electronic devices.

Rip 'n' read A hasty presentation of news by means of tearing copy directly from a wire machine and reading it on the air.

Roll cue A transitional statement that enables a performer to yield his or her presentation to another person or segment.

RTNDA An abbreviation for Radio Television News Directors Association, a group devoted to upgrading standards for news.

Run down An outline of a segment or program. Also called a *fact sheet*.

Screen Actors Guild (SAG) A union referred to as "sag."

Scale Minimum union wages.

Scene A portion of a drama or documentary, typically the moment of entrance to exit of a performer from a camera's field of view or microphone pickup.

Set The scenery (walls, landscapes, panoramas) for a production. Also called the *setting*.

Share A give-and-take relationship between performers, as in an interview or documentary. Also, a part of a viewing audience.

Shot A visual unit of a program or film during which an action occurs without interrupting the physical continuity by cutting or editing.

Sides Typed copies of a performer's part, including only his or her speeches, stage directions, and cues.

Soft news A news story having high human interest value, such as humor, novelty, gossip, and the like.

Sound bite A brief audio-video insert in a news report; often about 10 seconds used for attribution.

Source The original event; a person providing information for a story; equipment picking up audio-visual information.

Spot A commercial or public service announcement. Also, a brilliant white light emphasizing the main performer. A place in the organization of a program is referred to as a *spot*.

Stand by A verbal or visual cue given just prior to a rehearsal or on-air program.

Stand-in A substitute for the actual performer used to check lights and final preparations.

Straight A role similar to a performer's natural characteristics and age.

Studio A specially equipped room in which a performer rehearses or presents his or her act.

Summary A prepared statement with which a moderator closes a program.

Talent A collective term for all performers. Talent implies that the performer has exceptional ability.

Tag line A performer's speech or final line at the close of a story, segment, or program.

Talk A radio or television format consisting of information, news, discussion, and interviews.

Tally lights Tiny lights atop a television camera that indicate the camera is on the air.

Tape Any of several strips of plastic cut in various gauges and coated for different purposes such as recording sounds and picture, filling blanks, mending, and providing leader. *To tape* means to record.

Teaser A short enticement at the beginning of a program; a preview. In news, it consists of headlines; in drama, it may be a dramatic highlight presented in or out of story context.

Throw away To underplay a line or scene.

Time A term that means the segment is over. *Running time* is the duration of any program or segment, often measured in seconds. *Timing* refers to a performer's judgment in the delivery of material to receive maximum impact on an audience.

Track The portion of a film strip or tape that carries the sound impulse. Sophisticated sound uses multitracks.

TV-Q A measurement of audience response to a performer in terms of whether a performer is known and liked.

Unidirectional A narrow microphone pickup pattern projecting in a single direction and designed for a single performer.

Voice-over (VO) A performer reading copy over silent tape or film.

Vocal quality The audible characteristics of a voice including all of its nuances. A voice displaying a broad range of clear tones is considered to have high quality. A voice having excessive or unpleasing nasal, breathy, harsh, or strident characteristics has poor quality.

Voicer A prepared presentation by an identified reporter.

Wind screen A sponge rubber or latex sound-absorbing shield wrapped over a microphone to reduce extraneous noise.

Wind up A signal to end a segment, commonly the final 30 seconds. Also called *wrap-up*.

Wing it A term for doing a program without a rehearsal.

Wraparound A news story around which a reporter appears for a lead in and a final statement.